X

THE EMERGENCE OF NUMBER

THE EMERGENCE OF NUMBER

by J N Crossley

Monash University
Melbourne, Australia

World Scientific
Singapore • New Jersey • Hong Kong

Published by

World Scientific Publishing Co. Pte. Ltd.
P.O. Box 128, Farrer Road, Singapore 9128

U. S. A. office: World Scientific Publishing Co., Inc.
687 Hartwell Street, Teaneck NJ 07666, USA

THE EMERGENCE OF NUMBER

ISBN 9971-50-413-8
     9971-50-414-6 pbk

Printed in Singapore by General Printing and Publishing Services Pte. Ltd.

道可道非常道
名可名非常名

留兵綠道德經

Preface to the Second Edition

More than ten years have elapsed since I started work on the first edition. In that time support has grown for a more human approach to the development of mathematical concepts (see e.g. Davis and Hersh [1981]). The basic picture has not changed but the whole book has been revised, new material incorporated and substantial changes have been made (particularly to chapters III, V and VI). I believe part 3 (on real numbers) now presents a much clearer view of the subtle difficulties of the subject and the underlying assumptions which are largely unacknowledged by working mathematicians.

Perhaps the most distinctive feature of this book is its attempt to allow the modern reader to see through the eyes of the mathematicians of the past. There are many books on the history of mathematics which put the ancient mathematics into modern form but this hides all the insights of the ancients and usually does not show where the difficulties lay. Therefore we used the device of quoting the original authors as far as possible in the first edition and this approach has been retained. It is our hope that the reader will thereby be encouraged to seek out the original works and understand what the authors themselves meant, rather than rely on second hand paraphrases put out by later authors usually from a different culture and certainly with different aims.

The book remains an attempt to understand the workers of the past in their own terms and to see whence our present views came.

I should like to thank my colleague John Stillwell for his comments and proof-reading, Anne-Marie Vandenberg for again typing the manuscript so beautifully, Jim Mauldon and Tim Brook; Lencie Harding, the Inter-Library Loans Officer of the Hargrave Library, Monash, the Library of the Victoria University of Wellington and the kind reviewers of the first edition. Amongst these latter I should like to put on records my thanks in particular to Professor R.L. Wilder, now sadly deceased.

The advice of D.J. Whiteside recorded in the preface to the first edition still stands. It is reinforced by Halmos's prescription (in Halmos [1986]) for the scholar:

He knows the literature, and he trusts nobody: he himself examines the original paper. He acquires firsthand knowledge ... .

John N. Crossley

Monash University
17 August 1987

Preface to the First Edition

In this book I try to trace the origins and early development of three kinds of number: the ordinary, natural, counting numbers, complex and imaginary numbers and irrational numbers.

Numbers are so familiar to us that we rarely think of how they arose. In tracing their origins I have found it necessary to cover developments which have taken place over the last 12,000 years or so. Inevitably this has meant that I have been eclectic. It is my hope that the gentle reader will be moved to supplement my materials with his own discoveries and to test his own ideas and theories against such facts as are available. My aim has not been to present a theory of the development of number but to seek out its genesis. So the answers to the questions of how did particular kinds of number emerge are sometimes psychological, sometimes historical and only sometimes mathematical.

Many people and institutions have helped or advised me. They include Monash University (in particular its Hargrave Library staff), the Bodleian Library (specially the staff of Duke Humfrey and James Flavell), the Codrington Library (and its encyclopaedic librarian John Simmons), the State Library of Victoria, Lincoln College Library, Oxford, the departments of mathematics and philosophy of the University of the Philippines, Ateneo de Manila University, Lionel Glassey, Nigel Wilson, Bruce Barker-Benfield, Rupert Bruce-Mitford, Ivor Grattan-Guinness, Robin Gandy, John Mayberry, David Fowler, Cecily Tanner, Mordechai Feingold, Garry Tee, Richard Rouse, Michael Dummett, Michael Swift, Roger Highfield, Gale Sieveking, Stella Crossley, Robert M.W. Dixon, Peter Burke, John Sparrow, Ron Keightley, Alan Henry, Chris Ash, Charles Johnson and last, and most of all, my colleague Gordon Smith.

I am also grateful to Charles Thomas and the 1979 British Mathematical Colloquium for the opportunity to present an earlier version of chapters I and II, the Editor of the Gazette of the Australian Mathematical Society, where an earlier version of Chapter V appeared (see Crossley [1977,1978] in the references) and the Singapore Mathematical Society before whom an earlier version of chapters III and IV was read.

I am grateful to Eileen Badrian for some of the typing and especially to my secretary Anne-Marie Vandenberg who has been a constant help throughout.

In another practical way study leave from Monash and a Visiting Fellowship which allowed me to return to All Souls in 1979 were of great assistance.

I have had much advice. I have not always taken it though I have, I trust, always been grateful for it. One particular piece of advice was given me by D.J. Whiteside in Cambridge in 1975. It is to go and look at the originals. There is no better advice.

John N. Crossley

Monash University
26 March 1980

Contents

Chapter IV. Revelation

Part 3. Real Numbers

Chapter V. Irrationals

Chapter VI. The Totality of Real Numbers

To
Cressida

Prologue

That numbers have a timeless status is a view held both by most mathematicians and the world at large. It is a view that we shall challenge. While everyone would agree that talking about numbers depends on the state of our knowledge, it will also appear in the course of this book that what a number is and what numbers there are at any given point in history also depend on the state of knowledge at that time and on what human beings have done. We shall try to put numbers in an appropriate perspective and this perspective is a new one. We shall show that, far from being static, the concepts of number have continually developed from their earliest beginnings.

Even the simplest of numbers, the natural numbers 1, 2, 3, ... will be seen to have emerged only slowly into the abstraction we have today. Perhaps natural numbers were called by that name because they appeared to be practically innate. Certainly the view that they are innate has been held by distinguished anthropologists since the beginning of the science of anthropology (see e.g. Tylor [1871], vol. I, p. 221). On reflection they appear in a very different light as a phenomenon which has slowly grown and developed as the need has arisen. We shall see in Part 1 how they have developed from quite concrete beginnings into the sophisticated abstract concept we have today. We shall also see how the context in which they developed influenced that development.

It is possible to distinguish quite a number of stages and perhaps later workers may distinguish finer gradations in the process. Already one can see that they are not innate, though they are cultural universals (see Wilder [1973], p.33f.), they are present to a greater or lesser extent in all cultures. In our present culture they are highly abstract objects described by axioms but the move from concreteness to abstraction seems to have been a sudden one which took place due to Dedekind in the latter half of the nineteenth century.

In Part 2 we consider complex numbers: $\sqrt{-1}$ for example. These erupted on to the scene in the solution of cubic equations, though nowadays they are first encountered when trying to solve quadratic equations. We shall see how they received a pragmatic justification. We shall also show why they were not introduced before the sixteenth century.

Our final chapters, Part 3, are devoted to the real numbers. In particular we consider the introduction of irrational numbers and the gradual study of more and more such numbers until the whole totality of real numbers began to be studied in the late nineteenth century. We shall see the precise rôle of infinite processes in the description of real numbers. We shall also see that what appears intuitively obvious actually requires a new axiom and that the connexion between arithmetic and geometry which concerned the ancient Greeks has not yet been completely resolved.

Part 1

The Natural Numbers

It is probably familiar enough to most readers that many savage tribes are really unable to count, or at least have no numerals, above 2 or 3 or 4, and express all higher numbers by a word meaning "heap" or "plenty" (Gow [1884], p.4).

The numbers are free creations of man's mind, they serve as a means of apprehending the difference of things more easily and more sharply (Dedekind [1968], vol. 3, p.335).[1]

God made the integers, all the rest is men's work (Kronecker).[2]

The number which belongs to the concept F is the extension of the concept "equal to the concept F" (Frege [1884], p.84).[3]

These quotations are from the 1880's. They reflect the diversity of interest in, and ideas about, the natural numbers (1, 2, 3, etc.)[4] current in the late nineteenth century. In this chapter we shall consider the development of the natural numbers from Palaeolithic times (about ten thousand years ago). We shall principally be concerned with the development of the idea of (natural) number and we shall, as far as possible, follow that development in chronological order. To put it briefly, we shall try to show how the numbers developed, first in primitive societies and ultimately in the hands of mathematicians during the last hundred years.

We begin with the development of counting in primitive societies. For this we shall consider both archaeological and anthropological evidence. In looking at the latter, we shall make a great deal of use of linguistic evidence; that is, evidence obtainable from languages which possess different facilities for treating numbers from English. As we shall see, all languages have a limit to their counting words and it happens that, in general, other languages do not have vocabularies which enable one to count as far as in English. Having seen that there is a limit, a particular concern of ours will be with the slow advance of this limit to counting as time passes. As an example of this we shall exhibit a "primitive" counting system which is still in the process of developing.

Next we shall turn to the idea of larger and larger leaps in counting – for example, the way we go from ten to a hundred to a thousand to a million rather than counting in, say, tens all the time. Somewhere around this point the idea of *unending* repetition - of continually adding one (or ten) - emerges and with it the notion of the infinitude of (natural) numbers enters.   At first this is hidden in the innocuous-looking phrase "and so on".   Such informality is not precise enough and the move towards formality brings us to the end of the first chapter.

In the second chapter we trace, in a way which is of necessity somewhat eclectic, the emergence and development of the treatment of natural numbers by mathematicians.   In particular we shall watch the increasing precision of ideas as they come closer and closer to what we now call the principle of mathematical induction:

if the number 1 has a certain property, and if, whenever n has that property, so does the (next) number n+1, then every number possesses the property.

We view the crystallization of this notion by Dedekind and Peano as marking a profound distinction between two notions of natural number: one concrete and practical, the other abstract and belonging to mathematical theory.

We shall argue that this formalization of the idea of induction led, in the nineteenth century, to the replacement of intuitive notions of natural number, by a formal one.   We shall briefly consider the consequences of this in the early twentieth century, particularly Gödel's incompleteness theorem, and we shall end by showing how attempts are now being made to go beyond Gödel's theorem.   Whether this heralds a return to a more intuitive notion or rather just one further step in the historical process of making mathematical ideas more precise will only be determined in the years to come.   I believe, unlike, for example, Frege, that our present age is no more in possession of the ultimate characterization of number and numbers than any previous age.   It is this unending search for ultimate clarity and perfection which keeps mathematics, and indeed all intellectual life, alive.

Counting and the natural numbers are, to me, manifestations of certain developments in the human race.   I believe those developments were much more gradual and much less obvious than is usually supposed.   Although I shall dwell on certain features and certain writings, these should be regarded as merely representative: marking the way rather than being the only true guides.   They should be regarded more as checkpoints on a route rather than as the only interesting landforms.

Chapter I

Genesis

1 The start of the search

In a history of the London Missionary Society we read:

In the last decade of the eighteenth century the vast majority of the inhabitants of Great Britain knew less about India than those of today know about Patagonia, and their interest in the welfare of its myriad peoples was slighter far than their knowledge of the country (Lovett [1899], p.3).

The nineteenth century saw a tremendous upsurge in interest in primitive societies. The end of the eighteenth century had seen Captain Cook's landing in Australia. The next few years saw the beginning of missionary societies (cf. the quotation above) and a tremendous increase in exploration generally. Moreover, about the middle of the century anthropology and ethnography became firmly established[1]. Explorers and missionaries brought back fascinating accounts of primitive peoples, although these were not always totally reliable, as we shall see. We shall shortly turn to some of these accounts and later ones, but quite recently evidence has been produced for a form of counting or tallying going back to Upper Palaeolithic times, that is to say to about 20 000 to 10 000 B.C.. How much further back it will be possible to go as research advances, no-one, of course, knows.

At once we encounter the problem of what one means by counting and numbers (or number). Any definition we might give would not suffice for the whole of our considerations.[2] More importantly perhaps, some distinctions we are now able to make could not be made in earlier times. As a simple example, anyone writing in Latin would have found it difficult to express the difference between the ideas of "number" and "a number": both would be rendered by the same Latin word without "a".

The Shorter Oxford English Dictionary defines "to count" as "to tell over one by one, so as to ascertain the number of individuals in a collection; to number; to reckon up; also, to repeat the numerals one, two, three, etc., as *to c[ount] ten*". We wish to consider an earlier stage than this counting. If we consider the sorts of situation in which counting occurs we may distinguish three major types. First, as the external world appears to be made of relatively permanent, discrete objects or events, one may wish to consider a collection of such objects or events. Example: there are three hats. Second, one may associate internal or superficial aspects of oneself with such collections. Example: counting on one's fingers. Third, an abstract idea, or relatively abstract idea, of number may be communicable. Example: number systems such as the decimal system.

4

There are, as we shall endeavour to show, no hard and sharp boundaries between these three and our choice of these is mainly to help the reader through the diverse material which follows.

2. Palaeolithic counting

Our approach would not have been very different from that of Gow [1884] (see the first quotation at the beginning of this part) or many other nineteenth-century authors, had it not been for a long analysis by A. Marshack [1972]. In his recent studies he looked at markings on bones and stones which seem to indicate that in the Upper Palaeolithic period (*c.* 20 000-7 000 B.C.) men made such scratches systematically. The scratches were, he claims, arranged in series and there were gaps at certain points. His counting showed that the major groupings consisted of 29 or 30 marks, sometimes with further division into 7 or 14 (see e.g. *ibid.*, p.84, 93). Moreover, the marks were usually not all made with the same implement. This suggests that they were made at different times. Marshack favours the view that the markings correspond to the phases of the moon and that this is signalled by the grouping into (approximately) sevens (*ibid.*, *passim*, but see e.g. p.40, 100). We shall shortly consider how plausible Marshack's conclusions are, but the fact that the marks appear to have been made at different times – and sequentially – is important for us. It indicates that Palaeolithic Man was establishing a correlation between something (presumably external) and marks he was making, sequentially, on bone or stone. In order to do this he did not need number *words*: the marks on the stone would have been available as a visible indication if he wished to communicate with others.

Marshack first published his findings in his memoir [1970], basing his findings on only 6 objects. In a critical review Rosenfeld [1971] writes of this work:

Marshack proposes that a group of marks which share the same cross sections and kinetic properties [viz. orientation, depth and angle of the cutting tool] were made consecutively by one hand at one time. However, he goes on to say that the number of marks within any such group is significant and must have a quantitative notational meaning ... The problem, therefore, must be to assess whether these sequences of numbers [of marks] follow some identifiable pattern, or whether their arrangement is merely haphazard. Marshack proposes to test this with a model of the lunar month. His reasons for the choice of this particular model are not clear, except presumably that the author is satisfied with the fit (Rosenfeld [1971], p.318).

However, from Marshack's few examples his reviewer finds that

it becomes increasingly difficult to find any consistency in the manner in which the analysed groupings may be matched to the lunar model. One is forced to conclude that the few examples of Palaeolithic mobiliary art described in this book furnish little evidence of any form of systematic notation (ibid., p.319).

But Rosenfeld does not question the assertion that some sort of counting is going on.   After Marshack's glossy book appeared in 1972, another reviewer, A. Sieveking, wrote:

Micro-photography reveals that the marks on these pieces have, in a number of cases, been cut with a series of different tools, presumably at different times ... This sequential notation is a fact that is new to us, quite unsuspected and fascinating.   Of course such a discovery begs for explanation ... [and] I think the theory [of Marshack that the marks constitute a lunar calendar] is sound and that by far the most likely interpretation of these carefully notched tallies is that they are, as Marshack claims, a calendric record of some sort (Sieveking [1972], p.329).

In one case Marshack has found two eagle bones which duplicate each other, thereby suggesting that these were indeed communicable records (Marshack [1972], p.159-170).   After discussing one bone, Marshack continues:

The first plate of [Edouard Piette's L'Art *pendant l'age du Renne, Paris 1907][3] contains two small drawings of an engraved eagle bone from Le Placard I had just analyzed.   It looks like the same bone, but a careful comparison indicated that it was not.   It is a "sister bone," with similar multiple angles on the main face ... It may have been made by the same hand, certainly it was made in the same tradition by the same culture.   Two matching finds such as this must be extremely rare and I know of no other like them in the Upper Palaeolithic.*

At this stage, therefore, it appears that Palaeolithic Man was able to represent numbers occurring in some context (perhaps that of the lunar phases).   In this sense we might well choose to say that Palaeolithic Man counted, in a way, in at least one context.[4,5]   The crucial points about the markings are typified by Marshack's comments about one particular fish bone:

[A] schematic rendering reveals that the markings were intentional, sequential and carefully structured into groups (ibid., p.349-350).

A further development of tallying is nicely recorded in Mycenaean Greece.   Chadwick writes:

We find tablets used for rough working, and one revealing tablet (PY [LOS] Ea 59) has on its back a clear example of how addition sums were sometimes done; single strokes were written for each unit, and these were arranged in groups of ten, each ten being written as two fives one above the other (Chadwick [1973], p.32).

Of course tallying persists to this day.[6]

3. Is counting instinctive?

From what we have just seen, it appears that some form of counting was present a very long way back in man's prehistory.   One is tempted to ask whether counting is instinctive as Tylor, the early anthropologist, suggests in his fundamental work [1871].   Tylor quotes with approval a famous deaf and dumb person, Massieu, as asserting:

"I knew the numbers before my instruction, my fingers had taught me them. I did not know the cyphers; I counted on my fingers, and when the number passed 10 I made notches on a bit of wood" (Tylor [1871], vol. 1, p.221).

Tylor adds:

It is thus that all savage tribes have been taught arithmetic by their fingers (ibid., p.244).

But is it? It is possible that *after* primitive men have learnt to count they then do arithmetic using their fingers. But how do they learn to count? Indeed, do they count instinctively?

Let us consider first the question as to whether animals count at all and then look at a specific example of a primitive people. We shall see that having been introduced to the idea of counting, some animals can then, in a certain sense, count, while some primitive tribes rapidly adapt to counting though they may not have counted before.

W. Thorpe writes:

How is it that the perception of a certain number of eggs [in a nest] can influence the physiological activity of a bird? It is clear that such behaviour does not necessarily involve counting in any ordinary human sense of the term. The bird might merely be reacting to a certain visually observed proportion of "eggs to nest", or it might in some cases be reacting merely to the amount of stimulation received by the brooding surfaces. We are still very far from having satisfactory experimental evidence on this point. But consideration of this eggs-in-the-nest problem shows how careful we have to be in deciding whether or not a bird can count, and how complicated the idea of counting really is (Thorpe [1966], p.385).

That the experimental situation is very difficult is demonstrated by the story of Clever Hans, a horse allegedly able to solve arithmetic problems. When asked an arithmetic question, this horse would tap with its hoof until the correct number was reached. There was no hint of fraud by its trainer and it was only when the trainer was asked to put to the horse questions to which he himself did not have the answer that an explanation was forthcoming. The explanation is that the horse responded to involuntary and unconscious movements of his questioner's body. The trainer had no idea how the horse managed to answer the problems correctly (Katz [1953], p.13-16).

Using apparatus designed to avoid such inadvertent cues, Koehler has described how

the birds in the experiments of [Koehler] and his associates learnt what he calls "unnamed numbers" ... Koehler puts forward the hypothesis that man would never have started counting (i.e. to name numbers) without two pre-linguistic abilities which he has in common with birds. The first of these abilities is that of being able to compare groups of units presented simultaneously side by side by means of seen numbers of those units only, excluding all other clues ... The second ability is to estimate (i.e. to remember) numbers of incidents following each other, and thus to keep in mind numbers presented successively in time, independent of rhythm or any other clue which might be helpful.[8]

In fact a raven and a grey parrot managed to distinguish between boxes with 2, 3, 4, 5 or 6 spots on the lid, thus supporting the first hypothesis and a jackdaw and a pigeon managed to take exactly the required number of baits (up to 5) (Thorpe [1966], p.390).

However, this training has only worked with small, specified numbers.

*[I]t has not been possible yet to get the birds to generalize [to an arbitrary given number] and so provide a **general solution** of the problem [of transferring a number from one quality to another] (ibid., p.392).*

(We shall return to this problem of the *limits* of counting later.)

Thorpe concludes:

Such results as Koehler's[9] *are of the greatest interest psychologically in that they show some capabilities of birds which are apparently lying dormant and which natural selection may conceivably be able to develop further in the future (ibid., p.394).*

In fact these propensities appear sometimes to lie dormant in humans, too. In his report of 1908 Roth, writing rather in the style of the late nineteenth century writers with which we opened this chapter, says of some inhabitants of Queensland, Australia:

Natives do not possess special terms to express numbers over three collectively, everything beyond this being relatively either few or many. Not that they lack the mental ability to appreciate a conception of higher values - I have known of black children working at decimal fractions, and a young full-blood engaged as draughtsman in a large engineering works - but that the opportunity so seldom arises of having to exercise it. Ask a black as to the number of occupants in a camp, he will probably tell you there are few or many, and if pressed for further information, will mention the names of Tom, Dick, Harry, etc., ticking them off or not on either his fingers or in the sand, but always in pairs.[10] *He apparently takes a concrete view of the case, leaving you to form a mental picture of the number as a whole. He can certainly form such a mental picture for himself, because he will describe any large number of strangers, a flock of pigeons, anything in fact, of which the components are not individually known to him, in some such form as "plenty sit down all round about". (Roth [1908], p.79-80).*[11]

This sort of account is oft-repeated.

A modern example is given by Hale ([1975], p.1-3). He also points out the difference between counting words and quantification (by, as he calls them, "indefinite determiners"). Hale writes:

The Walbiri language, of Central Australia, has a set of four indefinite determiners:

8

/tjinta/	"singular one"
/tjirama/	"dual, two"
/wirkadu, mankurpa/	"paucal, several"
/panu/	"plural, many".

These correspond exactly to the four grammatical numbers distinguished in definite determiners, as exemplified by the following: /njampu/ "this, singular"; /njamputjara/ "these, dual"; /njampupatu/ "these, paucal"; and /njampura/ "these, plural". The indefinite paradigm functions in the language as a system of determiners - it is not a system of numerals, contrary to what one might be led to believe from the literature on Australia which sometimes identifies languages as having the numerals **one, two, three** *and* **many**. *The fact is, the indefinite determiner paradigm, as a whole, is not used in counting in Walbiri, any more than are the various definite determiner paradigms. What is true of Walbiri in this regard is, so far as I can tell, true of the other Australian languages which I have any knowledge of. Furthermore, in the Walbiri case at least, I believe that it would be correct to say that there is no single linguistic convention which is employed in situations in which the activity of counting, or exact enumeration, is a practical necessity (although there is a Walbiri-based method of exact enumeration which is highly favored, quite apart from the recent and not universally known, English-derived system of numerals[12] /wani, tuwu, tjiriyi, puwa, payipi, tjikitji, tjipini, yayiti, nayini, tini, lipini, .../). While it would be essentially correct to say that Walbiri lacks conventionalized numerals, it would be incorrect to say that Walbiri speakers, irrespective of their knowledge of Anglo-European culture, lack methods of exact enumeration. The indefinite determiner paradigm includes two members referring to exact numbers: /tjinta/ "singular" and /tjirama/, "dual".[13,14]*

There are in fact two sorts of evidence to suggest that the idea of number is present in a non-vocal form: one is the use of alternative descriptions like the above; the other is the facility with which words for higher numbers are adopted. We shall return to this second point later. Hale views aboriginal counting as follows:

I think that the correct way to understand the Australian counting system is as follows: conventionalized counting systems, i.e. numerals, are for the most part lacking, but counting itself is not lacking, in the sense that the principle of addition which underlies the activity of exact enumeration is everywhere present. In fact, I would like to argue that counting, in this sense, is universal, and whether or not a conventionalized inventory of numerals exists in a given language depends upon the extent to which exact enumeration[15] is of practical use or necessity to the people who speak the language. One might look upon the Walbiri lack of conventionalized numerals as a gap in the inventory of cultural items - since the principle which underlies counting is present, filling the gap is a rather trivial matter.[16] This view is entirely compatible with the observation that the English counting system is almost instantaneously mastered by Walbiris[17], who enter into situations where the use of money is important (quite independently of formal Western-style education, incidentally) (Hale [1975], p.3).[18]

9

Thus it would appear that the idea of counting lies dormant until evoked.

It will be useful to consider, for a while, an illuminating parallel to this development of counting in the development of words for colours. There seems to be a parallel between the gradual refinement of colour words and that of number words. As children we learnt the colours of the rainbow, though perhaps in different versions: some include just red, orange, yellow, green, blue and violet, while others add indigo. However, the distinction of six (or seven) colours does not appear always to have been universal. An extensive study of the names for colours by Berlin and Kay [1969] using a colour chart with 329 colour chips has shown that although any individual will recognize up to eleven basic colours – that is, the quality of each of the eleven colours is quite precisely determined and appears to depend purely on physiology – nevertheless the words which have developed appear to follow a definite pattern. Thus any language has basic words for from two to eleven of the colours. (Derivative words, such as cerise, are derived from other sources, in this case a fruit.) Languages exist which possess any initial part of the following list (*ibid*, p.14-15).

Black and White	Red	Green or Yellow	Yellow or Green	Blue	Brown	Any of	Purple Pink Orange Grey

Thus Baganda, a language of Africa (Uganda, N.W. Lake Victoria) possesses only *eru* (light shades etc. corresponding to our white), *dagaru* (darker shades, black) and *mynfu* (reddish shades, pink, orange, brown, purple, corresponding to our red). On the other hand, Tzeltal, a Mexican language, has *sak* (white, *'ikh'* (black), *cah* (red), *yaš* (green) and *k'an* (yellow), while English has, of course, words for all eleven.

We even find Aristotle referring (in *Meteorologica* III, 4374-5) to the rainbow having three colours red, green and violet and he regards yellow as merely the contrast between red and green.

Berlin and Kay's work appears to show that languages develop more and more colour words according to the above scheme and *"color categorization is not random and the foci of basic color terms are similar in all languages"* ([1969], p.10). Thus, although human capacity or potential appears to be the same, the abstraction necessary to distinguish all eleven basic colours may not arise in any given society.

Now let us return to the development of number. Analogy with the example of colour words suggests that, to some extent at any rate, the idea of number may be latent and only gradually emerge and develop. That certainly accords with the views of the psychologist Thorpe quoted above. The experiments on birds involved transference of the number from one collection of objects to another. There is linguistic evidence to suggest that this transference was only slowly acquired.[19] To this we now turn.

4. The abstraction of number

Omnibus ex nihilo ducendis sufficit unum *(One suffices in order to get everything from nothing; Leibniz [1768], p.348).*

This is the motto Leibniz proposed when he discovered the system of binary numbers now used extensively in computing, that is to say, the process of enumerating the numbers as 1, 10, 11, 100, 101, 110, ... in the scale of two (dyadic scale) (see Zacher [1973], p.225 f.).

Leibniz's view is a very sophisticated one. When we look at primitive peoples and their languages, we find that we are a long way from such abstraction. Lévy-Bruhl uses Leibniz's motto as a heading to a section of his book [1951] and then continues:

Prelogical mentality, however, which has no abstract concepts at command, does not proceed thus [i.e. in the manner of Leibniz]. It does not distinctly separate the number from the objects numbered (Lévy-Bruhl [1951], p.219, Eng. trans. p.192).

This lack of separation is quite evident in some Mexican languages. Thus in Chontal, a Mayan language spoken in the State of Tabasco in southern Mexico, counting words consist of two parts. The first is a quantifier (answering "how many?"), thus *un-* or *um-* or *u-* for one, an, a; *ca¹-* or *ca-* for two; *uš-* or *yuš-* for three; *cən-* or *cəm-* or *cə-* for four; *ho¹-* or *ho-* for five and *wək-* or *wəh-* for six. There are no quantifiers corresponding to numbers above six and "the quantifier for six is recognised by older speakers but seldom used" (Keller [1955], p.259).

The second part is a classifier. Keller lists 78 classifiers from which we quote a couple of examples: *-tek* for plants and standing trees; for example, *ca¹ tek co*, two stalks of corn, *untek te¹*, a tree; *-kuc* for loads of things carried on the back with a head strap, thus *unkuc si¹*, one-load-of firewood, *unkuc šan*, one-load-of palm (*ibid.*, p.260).

There is another Mayan language, Chol, close to Chontal, where the number system is well-developed but nevertheless prefixes for the particular kind of object are still used. Chontal itself has number words only up to five or six (*ibid.*, p.259), while Chol has words for numbers up to 159 999 and yet this language displays similarities with the Chontal system.[20] Thus Aulie gives the following examples ([1957], p.282):

Chontal	Chol	
uncit te¹	*hunc'iht te¹*	one stick
cak'e wah	*ca¹k'ehl wah*	two tortillas
ušlip'	*ušlihk*	three pieces
		(as of cloth)
cemp'is bu¹u	*cemp'is bu¹ul*	four measures
		of beans

Yet another example is provided by Berlin [1968] who exhibits 528 different classifiers for different sorts of objects in Tzeltal, another Mexican language. Thus *hoy* is used for a human group, standing, arranged in a circle, *p'o⟩* for an animal group with no specification as to position or alignment (*ibid.*, p.58-59), while there are various classifiers for different sorts of pieces of wood, plants, etc. (*ibid.*, p.46-47).

Notice that here we have not simply a word or phrase reserved for a specific type of subject, but a more abstract classification. This may be regarded as an intermediate stage between our (English) system, where number words usually have no connotation of particular objects[21] and that exemplified by a language of Fiji mentioned by Codrington. Thus Codrington relates:

> There is not, as far as I am aware, in Melanesia any way of counting by pairs like the use in Polynesia.[22] In Fiji and the Solomon Islands there are collective Nouns signifying tens of things very arbitrarily chosen, neither the number nor the name of the thing being expressed. Thus in Florida[23] na kua is ten eggs, na banana is ten baskets of food. In Florida these words are in no case the same as those in Fiji, and they are not so numerous, but the same objects are often counted in this manner. In Florida ten canoes or ten puddings are **na gobi**, which in Fiji are respectively **a udundu**, and **a wai**[24]; in Florida **na paga** is either ten pigs, or ten birds, or ten fish, or ten opossums; in Fiji ten pigs are **a rara**, ten fowls **a soga**, ten fish **a bola**. There are many other words of the same kind, naming tens of cocoanuts, breadfruit, crabs, shell-fish, bunches of bananas, baskets of nuts. In Fiji **bola** is a hundred canoes, **koro** a hundred cocoanuts, **a selavo** a thousand cocoanuts. In Florida **parego** is a collective noun for ten of anything; in Bugotu **selage** is ten, **tutugu** twenty, things of any kind (Codrington [1885], p.241).

One more example comes from Dobritzhoffer [1784]. Writing of the Abipones of South America, Dobritzhoffer says that they only have names for 1, 2, 3, while the word for "emu's toes"[25] is used for 4 and the word for "a beautiful skin" (referring to a particular skin with 5 colours) is used for 5.[26] Here we have a simple case of number words just petering out.

5. The development of counting

Thus we see that there is a separation of the abstract number from the classes of objects counted. It is not at all clear that once a human has counted one collection of objects he can immediately count another collection of the same number of objects. In this he appears like Koehler's birds. Further, some number words are only used for specific objects or tasks. For example, counting using parts of the body is used in the Ipili language in the Enga province of Papua New Guinea in the ceremonial exchange of goods (Lean [1985], vol.10, p.55; see also the discussion in *ibid.*, vol.12, p.101).

An alternative source of counting has been suggested by Cassirer. He feels that counting does not depend, even primarily, on external objects. He writes:

The differentiation of numbers starts, like that of spatial relations, from the human body and its members, thence extending over the whole of the sensuous, intuitive world. Everywhere man's own body provides the model for the first primitive enumeration: the first "counting" consists merely in designating certain differences found in external objects, by transferring them, as it were, to the body of the counter and so making them visible. All numerical concepts, accordingly, are purely mimetic hand concepts or other body concepts before they become verbal concepts. The counting gesture does not serve as a mere accompaniment to an otherwise independent numeral, but fuses in a sense with its signification and substance ([1953], vol.1, p.229).

Cassirer goes on to give several examples. But Seidenberg disputes this, arguing:

Yet the facts indicate that finger counting is learned, just as counting is ... [I]t should be understood that the number gestures of savages are conventional forms: if 1 is indicated by raising the left little finger, say, then a raised thumb would, in this connection, have no meaning. This shows that the gesture language is learned in quite the same way that the vocal language is (Seidenberg [1960], p.258).

This is further evidence to support our view that counting is not an instinctive activity. However, once the idea of counting has emerged, then the idea of going on counting does not seem to lie far below the surface. Man does indeed seem to possess an easily developed potential to count a long way. In the quotation from Hale (above §3), we saw how English words were rapidly assimilated. In many languages the local language is very rapidly supplemented by words borrowed from other languages.

Over the past few years in Manila in the Philippines I have often heard number words from three languages (Pilipino, Spanish and English) being used. While most people speak Pilipino, it is apparent that the choice of which language is used for particular numbers is dependent on context and tradition. Thus, although the native (official) language is used on banknotes, for large numbers English words are used in conversation. These words were imported during the American dominance of the Philippines. For small numbers Pilipino words are used, while Spanish words have been taken into Pilipino for e.g. twenty. Similarly in Mexico we have:

Chol terms for numbers in the thousands [are] gradually giving way to the Spanish mil (1,000) with some speakers (Aulie [1957], p.282).[27]

When we look at societies less influenced by imported words, then it is easier to locate the limits of precise number words. Writing shortly after 1900, Skeat and Blagden report of Malaya:

It has often been remarked that the purer dialects of the centre of the [Malay] Peninsula do not possess any native numerals for higher numbers than "three" (Skeat & Blagden [1906], vol.II, p.455).[28]

and also Pater Schebesta writes of some Malays:

The Jahai do not count above one, which they call "Nai". In so far as there are any other numerals, they are taken from Malay (Schebesta [1928], p.94).

13

Nevertheless, when people needed to count further there seemed to be no conceptual difficulty. The non-verbalized *ideas* of particular numbers appear to be present long before they may be needed in a verbal form and once counting is established there seems little difficulty in advancing rapidly.

One major development which is present in all verbal counting systems is the use of a single word or phrase which collects a group of objects together so that the number of groups can be counted. The simplest example is using a word like "hand" for "five" as "*lima*" in many Malayo-Polynesian languages (Codrington [1885], p.235 f.; also Lean [1985] *passim*). However, there seem to be different ways of counting and there is certainly no consistency in the size of the basic group as one goes through different languages. It should be noted, however, that the introduction of a word such as "hand" is not an immediate development.[29]

Marshack's work (see above, §2) suggests that the earliest counting of which we know was related to groupings of 7 or more. It therefore argues against the primacy of counting in twos or fives, which we shall discuss below (§6). It argues for a more concrete form of counting than counting using number words, a reification of counting.

6. Concrete Counting

We see what appears to be a similarly concrete form of counting in some primitive societies. For example, we have the following report from 1908 which gives three distinct ways of counting where, in two cases (tally-stick and fern-frond), there is a definite practical limit to the numbers counted and in the third ambiguity would render the hand-counting process useless. Note, however, that here the grouping word has not yet been reached.

> When a prun[30] is completed, the old men will arrange amongst themselves
> as to when and where the next is to be held, and a messenger will be sent to
> the different camps to tell them they are expected at such and such a place
> on a certain day. But many of the larger camps in the neighbourhood have
> their own calendar, which is similarly always being altered, so that the
> names of the days will, perhaps, not correspond. To obviate all mistakes,
> the messenger employs at least three methods to make matters clear - the
> message-stick, the fern-frond, and where it is understood, by mnemonics with
> the hand. The Yidinji blacks apparently used all three. The stick is
> employed by cutting nicks on it, each cut representing a day, but contrary
> to expectation I have never been able to find the passing of a day indicated
> by the crossing of a cut. With the fern-leaf, it is split in half, the
> number of leaflets left attached indicating the number of days of the
> interval, a leaflet being folded on itself for each day passed (Roth
> [1908], p.79-80).

Thus here we have what appears to be a nearly identical process to that reported by Marshack, supplemented by two other methods. The second, the fern-leaf, is similar though less permanent and possibly more prone to error, while easier to use. The third method depends more on the human element; indeed, it requires thought, but then it no longer requires any artifact.   Roth continues:

On the Lower Tully River[31] the intervening days are borne in mind by the different parts of the palm and digits, as follows, the messenger being able to explain to the various camps visited exactly how many days later they are expected:- Opening his left hand the reckoner names the first and second

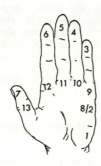

days as he points to the spots respectively so marked in the figure (Fig. 1), the same with the fourth, fifth and sixth; and now, with fingers all closed, he seizes the extended thumb and mentions the seventh, the date for which the next prun has been appointed.   But supposing that it has been decided to hold the next performance after an interval of thirteen instead of seven days, the reckoner will open his hand again and point respectively to the spots numbered eight to thirteen, the final day always ending with the thumb, giving them names identical with those already mentioned by him for the first to seventh day, thus:-

Fig. 1

1st day	=	chalgur
2nd or 8th day	=	chalguro-kabun
3rd or 9th day	=	meriri
4th or 10th day	=	mono-chano
5th or 11th day	=	moko-pulo
6th or 12th day	=	karapo
7th or 13th day	=	kari-unggol.

These words have no other significance, are absolutely distinct from the terms indicative of number, and are only applied to a day as a portion of the interval between the successive pruns, the idea of time-when being otherwise always reckoned by the numbers of sleeps.[32]   Amongst the blacks of the Upper Tully River the performance was held either on the eighth day or thirteenth day, the numbers referred to being shown in Fig. 2, where the names for the fourth and ninth, for the fifth and tenth, etc., are identical.   The Cairns Natives (the Yidinji), who had an interval of eleven days between the performances, puloga as they called them there, reckoned the intermediate days on both hands: first and second on ball and tip respectively of left thumb, third to sixth on tips of remaining fingers, the seventh to tenth on right hand fingers, commencing with the little one, the eleventh falling on right thumb; the names applied bore traces of the three numerals, as well as compounds of them (ibid., p.80).[33]

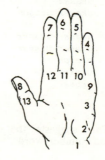

Fig.2

From what we have seen so far, it might appear that tallying, that is the marking of strokes on an object, is historically prior to counting by groups of, say, five or so.   Of course it is impossible to prove this

15

because the only type of evidence available for palaeolithic phenomena is artifacts, though we have seen that the marks on bone considered by Marshack do suggest grouping. However, another example of a primitive system (see below, §7) seems both to indicate that the co-existence of several modes of counting is more likely and also that there has been considerable interaction between the various modes leading to gradual refinement over the ages.

Thus we have one form of counting, the tally stick, which quite possibly has palaeolithic predecessors.[34] Counting on the fingers is also very primitive, though we cannot say how far back it goes. There are also very primitive forms of verbal counting. Roth mentions that the Aborigines he is discussing do possess some number words. However:

Natives do not possess special terms to express numbers over three collectively, everything beyond this being relatively either few or many ([ibid.], p.8).

We find this phenomenon is reported as being widespread. It also embraces the notion of grouping in groups of only two objects.

Over most of aboriginal Australia one finds essentially only two number words, 'one' and 'two': many of the tribes are reported definitely as not counting beyond 2, and as indicating higher multiplicities by the word "many", while others go a little further by compounding these words - for example, expressing 3 as "two-one", 4 as "two-two", and 5 as "two-two-one". A similar method of counting, the so-called 2-system, is found in New Guinea, in South America, and in South Africa ...

Gumulgal (Australia)	Bakairi (South America)	Bushman (South Africa)
1 urapon	tokale	xa
2 ukasar	ahage	t'oa
3 ukasar-urapon	ahage tokale (or ahewao)	'quo
4 ukasar-ukasar	ahage ahage	t'oa-t'oa
5 ukasar-ukasar-urapon	ahage ahage tokale	t'oa-t'oa-t'a
6 ukasar-ukasar-ukasar	ahage ahage ahage	t'oa-t'oa-t'oa

(Seidenberg [1960], p.216).

This suggests that men could count, in the sense of tallying, before they had number words.[35] Hale gives other examples from Australia.

Using these forms [/tjinta/ "singular" and /tjirama/ "dual"], with the principle of addition[36], it is possible to refer, with precision, to numbers higher than two: /tjiramakaṛi-tjinta/ "three", /tjiramakaṛi-tjiramakaṛi/ "four", /tjiramakaṛi-tjiramakaṛi-tjinta/ 'five', and so on. In principle, there is no upper limit to this, although as the numbers get higher, the corresponding names for them become longer and more impractical; and in some Australian languages (like Gunwinjgu of Western Arnhem Land, for instance), where counting is of greater practical importance than it is for the

Walbiri, shortcuts have been adopted[37], *and to some extent conventionalized, to convert the higher numerals into a more manageable form (e.g. the use of* hand *for* five, hand + hand *for* ten, hand+hand+foot *for* fifteen, *and so on) (Hale [1975], p.296).*

We should beware here of the idea that "in principle, there is no upper limit to" counting two-two-two-two-... for certainly the Aborigines do not *keep on* iterating two-two-... . Indeed, the idea of counting *indefinitely* far seems to be by no means primitive and we shall return to this point below (see §10). So far we have only been concerned with the way that the *abstract* idea of number has gradually developed from where it was impossible to speak of, for example, the number "two" without talking about two this or two that.

Just now we have been considering how the concept of number develops and expands, but before we consider the questions of the beginnings of counting and the limits of counting, we shall pause to consider a specific example of a primitive counting system.

7. A primitive system

We present a primitive system for a limited number of numbers. That is to say, a system without a systematic way of proceeding to arbitrarily high numbers. The example is very modern, indeed, so modern that it is still in the process of evolution! In fact it is a system which strictly speaking counts logarithmically, going up in multiples of 1 000. This corresponds (after taking logarithms) to counting by threes.

This system is the system of names for very large (and very small) multiples.[38] It has its main origins in electronics. Here the need arose to deal with units up to 10^{12} (that is, 1 000 000 000 000) and down to 10^{-18} (that is, 1/1 000 000 000 000 000 000). The English words (up to billion and down to trillion in British English) were not used because of the confusion between British and American English[39]. The point I wish to dwell on is the *development* of this system. The system consists of prefixes. We consider both up and down because the steps are, so far as we are concerned, the important thing.

In its present state, the system consists of the following prefixes:

kilo- for 1000, i.e. 10^3 times multiple.	*milli-* for 1/1000, i.e. 10^{-3} fraction
mega- for 10^6.	*micro-* for 10^{-6}
giga- for 10^9.	*nano-* for 10^{-9}
tera- for 10^{12}.	*pico-* for 10^{-12}
	femto- for 10^{-15}, and
	atto- for 10^{-18}.

(I have also seen *exa-* for 10^{15} and *peta-* for 10^{18}, but these do not appear to be in common use.)

The origins of these terms in chronological order are as follows.

17

kilo- 10^3 *An arbitrary derivative of [the Greek] χιλιοι, a thousand, introduced in French in 1795 at the institution of the metric system ... (O.E.D.)*

It is not until 1816 that we find milli-:

milli- 10^{-3} *combining form of [Latin] mille, thousand ,... [First in print in] 1816. P. Kelly, Metrology 17. The word Milli expresses the 1000th part. (O.E.D.)*

Gradually the system expands; thus we have

mega- 10^6 *[representing Greek] μεγα-, [combining] form of μέγας, great, ... [First in print in] 1868. L. Clark Electr[ical] Meas[urement] 43. One million ohms =1 megaohm. (O.E.D.)*

Not so long after we find:

micro- 10^{-6} *[representing Greek] μικρο- [combining] form of μικρός, small, used chiefly in scientific terms. [First in print in] 1873 Rep[ort] of the] Brit[ish] Assoc[iation] 224. For multiplication or division by a million, the prefixes* **mega** *and* **micro** *may conveniently be employed. (O.E.D. micro- 5a.)*

It was not until the twentieth century's rapid development of electronics that *nano, pico* and the other terms we consider came into use.

nano- 10^9 *[from] Latin [nan-us, [Greek] ναν-os, dwarf + o.] [First appeared in print] 1947. Compt[es] Rend[ues] de la 14me Con[férence de l'Union Internationale de Chimie] 1952 Wireless World May 187/2. The prefixes pico and nano became popular in this country [the United Kingdom] fifteen or twenty years ago, mostly through the technical publications of Philips and others, with pico as favourite. (O.E.D.)*

pico- 10^{-12}*[from] Sp[anish] pico = beak, peak, little bit. [For date, etc., see nano- above.]*

The remaining prefixes we are considering have been formally adopted by international committees, though there are some (*peta-* for 10^{15}, *exa-* for 10^{18}) whose application the author does not know, and *beva-* for 10^9 and *myria-* for 10^4 which are used only in the U.S.A. and France respectively in connexion with particular units. These latter units are not generally accepted (cf. Dresner [1971], Appendix 3, p.231).

giga- 10^9 *An arbitrary derivative of [Greek] γιγᾶς, giant, [First in print in] 1951, Symbols, Signs & Abbrev[iations]. (R[oyal] Soc[iety]) 15/1 Prefixed to abbreviations for the names of units indicating multiples. [Also in] 1968*

Nature 19 Oct[ober] 311/2 CIPM at its meeting in 1958 recommended the prefix giga ... for the multiple 10^9. (O.E.D.)

tera- 10^{12} I can find only in the Concise Oxford Dictionary [from Greek] *teras* [τεραs], monster.

femto- 10^{-15}*and atto-* 10^{-18} *come from the Danish and Norwegian* **femten** *[fifteen] and* **atten** *[eighteen]. [They first appear in print in] 1963 Nature, 16 Mar. 1056/1 Int[ernational] Comm[ittee on Weights and Measures] adopted two new prefixes ... femto, 10^{-15},... and atto, 10^{-18}, ... [and Nature provides a list from atto- to tera-]* (O.E.D. femto-)

In 1968 George Gamow [1968] suggested naming units after people. He proposed 1 Hubble = 10^9 light years and 1 Rutherford = 10^9 electron volts, but this drew an interesting response which shows there was no obvious way to expand the system. Denvert and Oakland wrote:

Is there some exotic language which provides readily voiced prefixes suggesting 15, 18 and so on with initial letters acceptable for use as abbreviated forms?[40]

As an interim and somewhat retrograde step pending international agreement on such further prefixes, we can revert to the use of compound prefixes: the teraterametre (TTm: 10^{24} metres), for example, would be unambiguous and could cope immediately with cosmic distances (Denvert & Oakland [1968], p.311).

We observe that this last suggestion is very reminiscent of the reduplication in aboriginal languages where one counts one, two, two-one, two-two, two-two-one, etc. So this suggestion is in fact to use a very primitive mode for continuing counting in our supposedly advanced age!

We suggest that the way this system of prefixes came into being - that is, by gradual steps over a century and a half - mirrors the way that primitive systems have developed in general. Basically, new terms were not introduced until there was a need for them and, throughout, there was no attempt at, and no idea of, a continuing sequence but only a pragmatic accretion of terms as need arose.

The salient features of the particular primitive system we are considering are:

(i) terms are added as need arises and there is no obvious stopping place, although

(ii) there is clearly (from our twentieth-century viewpoint) the *possibility* of continuing the sequence indefinitely and

(iii) there are attempts at a regular continuation.

Regarding (iii) we recall the derivations of *femto-* and *atto-* and also the plea of Denvert and Oakland above for an "exotic" language, not Latin, which is inadequate and over-used, which would allow a "natural" and indefinite continuation.[41]   We shall return to the idea of indefinite continuations more than once (see below, §§9 and 11).   Before doing this we pursue primitiveness as far as we can.

8. The smallest numbers

In the primitive system we have just presented, "one" is represented by "kilo-" (Greek), meaning "thousand" and thus is without mystery.   However, in our search for the beginnings of counting the situation is much less clear and much more mysterious.

In summing up his researches on the origin of counting, Seidenberg found

> ... that counting was frequently the central feature of a rite, and that participants in ritual were numbered.[42]   This suggested to us the hypothesis that counting was invented as a means of calling participants onto the ritual scene.   With this hypothesis, we examined ancient thought for support [and] found it (Seidenberg [1962], p.36).

Just how far one can penetrate psychologically into these ideas is a difficult problem, but it is clear they are deeply involved in the human psyche.   At the end of an excellent Jungian survey, von Franz writes:

> Among the many mathematical primary intuitions, or a priori ideas, the "natural numbers" seem psychologically the most interesting.   Not only do they serve our conscious everyday measuring and counting operations; they have for centuries been the only existing means for "reading" the meaning of such ancient forms of divination as astrology, numerology, geomancy, etc. - all of which have been investigated by Jung in terms of his theory of synchronicity.   Furthermore, the natural numbers - viewed from a psychological angle - must certainly be archetypal representations, for we are forced to think about them in certain definite ways.   Nobody, for instance, can deny that 2 is the only existing even primary[43] number, even if he has never thought about it consciously before.   In other words, numbers are not concepts consciously invented by men for purposes of calculation: They are spontaneous and autonomous products of the unconscious - as are other archetypal symbols.

> But the natural numbers are also qualities adherent to outer objects: We can assert and count that there are two stones here or three trees there. Even if we strip outer objects of all such qualities as color, temperature, size, etc., there still remains their "manyness" or special multiplicity. Yet these same numbers are also just as indisputably parts of our own mental set-up - abstract concepts that we can study without looking at outer objects.   Numbers thus appear to be a tangible connection between the spheres of matter and psyche.   According to hints dropped by Jung it is here that the most fruitful field of further investigation might be found (Jung et al. [1964], p.385-386).

Support for these ideas comes from many quarters. In particular, Seidenberg believes that "two" is very important because of the frequency of processions of couples in various rituals.

If we agree that counting is not a simple process invented in one stroke, we may examine it for elements it contains that could have had an independent existence and out of which it might have grown. There is no particular call to count the fingers, and we may safely suppose that they have no bearing on the invention. The basic things needed for counting are a definite sequence of words and a familiar activity in which they are employed. The creation ritual offers us precisely such sequence and activity. Processions of couples in ritual are well known. "Male and female he created them."[44] "There went in two and two unto Noah into the ark, the male and the female, as God had commanded Noah."[45] ... The sequence of words so used might have come to be used as the initial number-words (Seidenberg [1962], p.8).[46]

Certainly many (and varied) authors suggest closely related religious or mystical origins.

In the olden days there lived in the sky two gods, Kaptan and Maguayan. Kaptan fell in love with Maguayan and they were married. One day Kaptan and Maguayan quarreled, as many couples do after the honeymoon. In the fit of anger, Kaptan told his wife to go away. With heavy heart, Maguayan left.

When the goddess was gone, god Kaptan felt very, very lonely, ... So to while away his sorrows, the repentant god created the earth and planted a bamboo in a garden called kahilwayan. ... God Kaptan was watching the breeze play among the leaves of the bamboo; a thought came to him, and, before he realized what it was all about, he was murmuring to himself: "I will make creatures to take care of these plants for me." No sooner had he uttered these words than the bamboo broke into two sections. From the slit of the node stepped out the first man and woman (Jocano [1969], p.35-37).

In Southeast Asia, besides, we know that the spirits of trees are called Nagas, which are snake divinities. But these snakes are also the masters of rain. In the related mythology of India we know that they have yearly to be cut in two by Indra, the sky god, in order to free the monsoon from their all-containing bellies. For tribes in Guyana, this freeing of the waters was originally produced by cutting down the World Tree. What put these two images, the snake and the tree, in such close parallel is a fact of some psychological curiosity; they are symbolically bisexual.

... Where do babies come from? We must now introduce another metaphor having to do with beginnings, that of sexual generation. The Platonic doctrine - which is also the Hindu one - sees this as a unitary process requiring three distinct natures. Plato describes these three natures as "first, that which is in process of generation; secondly, that in which the generation takes place; and thirdly, that of which the thing generated is a resemblance. And we may liken the receiving principle to a mother, and the source or spring to a father, and the intermediate nature to a child."

This is a new metaphor, but it is only another way of seeing the productive circulation of the four elements. The primal mother and father are no other than the dragon in its bisexuality. So we now see the answer

21

to the problem of how to create something out of nothing - it is to cut into
two the self-copulating dragon and so release the child of matter from the
waters of its fertilizing birth (Huxley [1974], p.96 f.).

Certainly "one" has been a problem since the time of the Pythagoreans
when it was not even considered to be a number.

But he [Plato] agreed with the Pythagoreans in saying that the One is
substance and not a predicate of something else; and in saying that the
Numbers are the causes of the reality of other things he agreed with them
[the Pythagoreans] (Aristotle, Metaphysics, 987b).

Further, all the small numbers have long had a history very much
tangled up with mystical meanings.[47]   It is clear that much more research
is needed, presumably to be carried out by psychologists, psychiatrists and
anthropologists.   Equally evident is the fact that (small) numbers are
intimately bound up with the human spirit and are by no means only cold,
lifeless abstract objects.

We now turn to the other extreme and investigate the limits of
counting.

9. The limits of counting

There are at least three senses in which the phrase "limit of counting"
can be interpreted.   First, there is the practical limit determined by the
smallest number which has no name in a given language.[48]   Second, there is
the situation where there is no clear way to proceed and third, there is the
question of the indefinite continuation of counting, i.e. the case of there
being no limit.

Concerning the first meaning, that is, the existence of a number
without a name, we see from the primitive system of §7 that there is no
natural way to continue – indeed Denvert and Oakland [1968] were asking for
a "natural" way – and the system is not extended until the need arises for
an additional name.   Similarly, in Chol the system stops where there is no
name[49] for 20 × 20 × 20 × 20.   Again in Archimedes' Sand Reckoner (Heath
[1897], p.347) we find descriptions of numbers up to

$$A = 10^{8 \cdot 10^8}$$

but after that, Archimedes just starts again and proceeds up to A^2, A^3 and
so on.[50]   No general procedure for continuing indefinitely far is given and
indeed Archimedes was not seeking such.[51]

In a variety of languages, after a certain stage, it is possible to
continue indefinitely simply by repetition.   Thus in 2-counting, where we
usually find counting of the form one, two, two-one, two-two, two-two-one,
once the idea of repeating "one" and "two" alternately and adding an extra
"two" has been grasped, there is no limit - other than the time and patience

of speaker and listener.   This phenomenon occurs in English, Rumanian and elsewhere.[52]   In British English one can use million million and million million million for 1 000 000 000 000 and 1 000 000 000 000 000 000 and the process can be continued.   In 1494 Nicholas Chuquet introduced terminology and notation for such large numbers.   He marked off groups of six digits by dots.   So the first dot indicates a million, thus 1.000000.   Chuquet did not publish his work (see Marre [1880] and Flegg [1985]), but his pupil La Roche did (almost *verbatim*!) and La Roche continues:

> ... the second dot [indicates] billion, the third trillion, the fourth quadrillion, the fifth quillion, the sixth sixlion, the seventh septilion, the eighth octilion, the ninth nonillion.   And so on as far as one wishes to proceed (La Roche [1520], f.7r).[53]

He then gives as example

> 745324 trillions.   804300. billions.   700023.   millions.   654321.
> Exemple 745324430800700023654321.

Thereby including a mistake.   (Chuquet's manuscript, see Marre [1880], contains a different example.)

But is it true that one can go on as far as one wishes?   Evidently not : for the same reasons as in our modern primitive system.   There is no obvious way to continue giving different names beyond a million million blocks of six digits.   That is because Latin, like any other natural language, has only a finite list of words so the supply of Latin roots must run out.   In the same way English eventually stops having names for large numbers.   Adventurous writers such as Smorynski [*op.cit.*] do extend the process somewhat.   W.D. Johnstone [1975], besides giving a list of the [American] English words from billion (= 1 000 000 000) through decillion (= 10^{33}), goes up the Latinized scale to vigintillion (= 10^{63}), omits other such names, pausing at googol (= 10^{100}), jumps back to Latin for quintoquadragintillion (= 10^{138}; Latin quinque et quadraginta = 45 but quintus is fifth) and then goes fancifully to millimillion "devised by Rudolf Ondrejka" = 10 to the 6 [American] billion, ending up with "1 Skew Named for Stanley Skewes, professor at Cape Town University, South Africa, the unit equals 10 to the power 10 to the power 10 to the power 34."

The third interpretation, that is, whether it is possible to continue counting indefinitely, *we* know is acceptable, and we shall later look (§11) at how this idea developed.   This leaves the second, and most ambiguous, interpretation.   Again it splits and there are at least three subcases. First, a temporary or perhaps semi-permanent block to counting may exist; second, more than a certain number may be called "arbitrarily large" or "infinite" or some such; and third, the idea of overcoming a limit and repeating an action or reapplying a device may enter.   This last we shall deal with in the next section.

We turn to the first, the temporary resting place.   In the quotation from Codrington above (§ 4) we noted the occurrence of collective nouns for specific classes of objects, e.g. *na kua* for ten eggs.   Besides this, in counting we have noted that it often happens that a new word or concept is introduced which changes the pace.   This happens, for example, when one gets to 5-counting, that is, counting by fives rather than twos or again in going from counting in tens to counting in hundreds.   (We English speakers are most familiar with 10-counting when we use the words two, three, etc., somewhat elided, to form twenty, thirty, and thus proceed at 10 times the previous rate.)   However, the change in the pace of counting may be surprisingly gradual.   Thus Codrington quotes the Rev. J. Inglis as saying:

> *The Papuans proceed thus: They count the fingers on one hand till they come to five, and then they say my hand, whatever that word may be in the language, for five; then my hand and one for six, my hand and two for seven, and so on till they come to ten, for which they say my two hands; they do the same thing with their ten toes, and then say my two hands and my two feet for twenty.   All beyond this in Aneiyumese is many, a great many, a great great many (Codrington [1885], p.226, n.1).*[54]

In fact Inglis goes so far as not to "admit the native numerals into his Grammar of the Anaiteum language at all, and only the first four into his vocabulary" (*ibid.*), but Codrington also notes that the influence of missionaries has affected the language and English numerals have been substituted for native ones.   It therefore is sometimes difficult to determine where and when the native numerals stop.   The same difficulty is remarked on by Pater Schebesta (see above, §5).

Thus here it is difficult to ascertain where the system stopped before missionaries arrived.   We find similar modes of counting in South America.

> *[C]onsider again the Guarays.   These count: 5 = 'our hand is finished', 6 = 'one on the other hand', and so forth, to 10 = 'finished are both our hands': after which they continue with Spanish (Seidenberg [1960], p.245).*

Cassirer also points out that in the language of the Inuit, who he calls Eskimo, 20 is expressed by "a man is completed" (Cassirer [1953], p.230).   Cassirer also gives an example of several limit stages in the case of the Sotho (*ibid.*).   Here, the word for "five" means literally "complete the hand" and that for "six" means "jump", i.e. "jump to the other hand" (*ibid.*, p.230, n.83).   What readily springs to our minds is counting up to twenty by using both fingers and toes, but

> *These motions need not be limited to the hands and feet, the fingers and toes, but can extend to other parts of the body.   In British New Guinea, the sequence in counting runs from the fingers of the left hand to the wrist, the elbow, the shoulder, the left side of the neck, the left breast, the chest, the right breast, the right side of the neck, etc.; in other regions the shoulder, the clavicular hollow, the navel, the neck, or the nose, eye, and ear are used (Cassirer [1953], p.230).*[55]

Glen Lean ([1985], especially vol.10, p.55-58, vol.12, p.100-102) gives accounts of counting in many Papua New Guinea languages where body parts are used.   (While these involve unusual lengths, e.g. 47 in Kewa, 19 in Gidra, they also often involve the more familiar cycles of length two, five or ten.)

It is clear to us that in the case of the Inuit, we could continue counting by using "two men are complete" for 40 and then going on to 60, 80, etc..   Nevertheless, it is not clear what to do after 399, though one would expect mathematicians experienced with number systems to transfer those techniques to the Eskimo situation.[56]   The important point is that, every

24

now and again, a new idea or a new comprehensive term is required to push any system of numbers further. Of course, if there is no pressure to count further one cannot expect an appropriate term to be extant.[57]

10. Imprecise limits

The second notion of "limit of counting" we have touched on already, in particular in the comments of Hale above (§6) regarding the language of the Walbiri. This notion appears in a number of Australian languages. Thus of the Parnkalla language in South Australia we read:

Sing.	Dual	Plur.
Yurra, *man*	Yurralbelli, *two men*	yurrarri, *men.*

There is another form for the plural number which may be properly called the **intensive plural**, *as it is only used when a great number or quantity is to be expressed (Schürmann [1844], p.4).*

Codrington nicely demonstrates how names for large numbers become very vague and indefinite.

Thousand. - *As high numbers are reached there is no doubt an increasing vagueness in their application, yet there can be no doubt but that Melanesians count with accuracy thousands of bananas, yams, and cocoanuts for feasts. The indefiniteness is shown in the word* **tar**, *which in the Banks' Islands is used for a thousand and also for very many, the same being a hundred in Espiritu Santo. In Nengone to count a thousand was to go as far as could be reached, e* **dongo***, finish. The Fiji* **udolu**, *thousand (the same word as* **nol** *in* **mel nol** *above[58]), means all, complete. In Wano of San Cristoval they have no word for a thousand. There is a word in use in Florida and Bugotu,* **mola**, *which is used indefinitely for a great number beyond count; and this, but doubtfully, is given in Malanta and Ulawa for a thousand.*

To go accurately beyond a thousand is not commonly possible, except as two or three or so many thousand; if there be a word said to mean ten thousand a certain indefiniteness hangs about it. If the Malagasy **alina** *means ten thousand, the meaning of the word is still 'night',[59] and there is a certain absurdity in saying* **alina roa** *"two nights", for twenty thousand, using a word for a certain number which denies the possibility of counting. In the Banks' Islands* **tar mataqelaqela** *is literally "eye-blind thousand", many beyond count. Figurative expressions show how the unpractised mind fails to rise to exactness in high numbers. In Torres Islands they use* **dor paka** *banyan roots, for very many beyond count, at Vaturana* **rau na hai** *leaves ofatree; in Malanta they exclaim* **warehune huto!** *opossum's hairs!* **idumie one!** *count the sand! In Fiji, however, the name of a tally like* **vatu loa**, *a black stone, no doubt is used with a definite number in view, though a* *number so large as one hundred thousand is given, and while yet* **oba** *is said to be used indefinitely for a lower number as well as for ten thousand. In the same language* **vetelei, wokaniu,** *are given for a million (Codrington [1885], p.250-251).*

Finally in this section a story about the Tongans which, as Seidenberg says, "not only illustrates the natives' love of horseplay but shows that they are quite familiar with their numbers and ready to make mental constructions on the basis of them" (Seidenberg [1962], p.280). (This story also illustrates how important it is to use primary sources wherever possible, since extraneous and erroneous data is often introduced in the transmission of information.) Seidenberg goes on to quote (not totally accurately) Tylor's remarks which are in turn a bowdlerized version of the report of Mariner. Mariner's report was written up by Martin [1818]. He reports on the voyage by Dumont D'Urville (who is best known for his Pacific voyage in search of the missing French sailor La Pérouse) on which La Billardière was naturalist. La Billardière several times on this voyage collected vocabularies. On this occasion a ship became beached and he was apparently able to spend a lot of time talking to the Tongans because he has a quite extensive vocabulary for Tonga, or the Iles des Amis (Friendly Isles) as they were then called. His number word list starts off normally enough, including *nima* for five (signifying "hand"), the usual sort of word in this part of the world as we have already noticed. But the list goes up to 1 000 000 000 000 000. Mariner describes the Tongan system of number words and gives an example for 95 741. He then adds in a footnote:

Their capability of expressing such high numbers in this decimal mode appears to be suspected by some readers; but we ought to reflect, that a people who are in the frequent habit of counting out yams, &c. to the amount of one, two, or three thousand, must become tolerably good numerators, by finding out some method of rendering the task of counting more easy.[60]

After a few more detailed comments, Mariner (as reported by Martin) goes on:

It may appear strange that they have particular names for such high numbers as 10,000 and 100,000, **mano,** *and* **giloo,** *for they certainly have no use for them. They often have occasion to count yams to the number of a thousand, or more, and sometimes to the amount of two or three thousand, but never higher. M. Labillardière, however, has had the perseverance to interrogate the natives, and obtain particular names for numbers as high as 1,000,000,000,000,000 !! Here, however, he has overshot the mark, and instead of names of numbers, has only furnished us with names of things very remote from his speculations at that time: for 1,000,000 he gives us* **nanoo,** *which has no meaning that we can discover; for 10,000,000* **laoalai,** *which should be* **lööóle** *(according to our spelling), which means the praeputium; for 100,000,000* **laounoua** *(low noa), which means* **nonsense;** *1,000,000,000* **liaguee,** *which we take for* **liagi,** *and is the name of a game played with the hands, with which probably he made signs; for 10,000,000,000* **tolo tafai** *(tole ho fáë), for which see the Vocabulary*[61]*; 1,000,000,000,000,* **lingha** *(linga), see the Vocabulary*[62]*; for a higher number they give him* **nava** *(the glans penis); for a still higher number,* **kaimaau** *(ky ma ow), by which they tell him to eat up the things which they have just been naming to him; but M. Labillardière was not probably the first subject of this sort of Tonga wit, which is very common with them. In the other numbers he is tolerably* *correct, except in putting* **giloo** *for* **mano***, and* **mano***for* **giloo***: his general accuracy in respect to the numbers does him great credit (Martin [1818], p.370).*

In fact Houton la Billardière's own vocabulary ([1800], vol. ii vocabularies, p.55) includes *lao* meaning testicles. The other words noted by Mariner are in the same vocabulary (*op.cit.*, p.57) and are included in a complete list proceeding by ten-fold steps and ending with 1 000 000 000 000 000 and then infinity: "nombre infini, *oki*".

A later vocabulary by the commander of the expedition himself, Dumont d'Urville, stops with *mano* (10 000) and *guilon* (100 000) (Dumont d'Urville [1826-7], *Philologie*, p.56): this is information from an Englishman called Singleton who had lived in Tonga for twenty years. But Dumont d'Urville had noted the irony of the Tongans in his report of the voyage (*ibid.*, vol.IV, p.334-335):

Mariner has observed that [the Tongan people] frequently employ this kind of irony which consists of saying the opposite to what one wants to express the better to persuade the person one is addressing.[63]

Having seen how vagueness was used to deal with indefinitely large numbers, it will be our next concern to see how the construction of even larger numbers came to be continued and this leads us to the third interpretation mentioned above.[64]

11. Unending repetition

Lévy-Bruhl writes:

It is usually admitted as a natural fact, requiring no examination, that numeration starts with the unit, and that different numbers are formed by successive additions of units to each of the preceding numbers. This is, in fact, the most simple process, and the one which imposes itself upon logical thought when it becomes conscious of its functioning (Lévy-Bruhl [1951], Eng. tr. p.192).

Just how simple is this "most simple process"? And is not the last clause of the above quotation looking at things from the wrong end?

So far we have seen how gradual the development of ideas in counting has been. We have also seen how quickly words for larger numbers are absorbed by people who previously had no names for those numbers. So it is important to separate the *process* of adding an extra tally or ticking off one more object and the *idea* of, or the idea that there is, such a process. To move from the process to the idea of the process one has to go up one level of abstraction.

Already in Palaeolithic times the process is present (see above, §2), but even a couple of hundred, or fewer, years ago the idea of counting indefinitely was not current in many parts of the world.[65] We have seen above (§10) in Codrington's report on Florida and Bugotu that they have a word, *mola*, for indefinitely many.   Again, in Australia we find that, in the Chaap wuurong language, counting went as follows:

Four	...	puuliit baa puuliit	—	*two and two*
Five	...	Koep mun'ya	—	*one hand (outspread)*
Ten	...	Puuliit mun'ya	—	*two hands (outspread)*
Twenty	...	Koep mam	—	*one twenty*
Fifty	...	Puuliit mam, baa	—	*two twenties and*
		puuliit mun'ya	—	*two hands*
Ninety	...	Puuliit mam, puuliit	—	*two twenties, two twenties*
		mam, baa puuliit mun'ya	—	*and two hands*

and yet we have

One hundred[66] ... Larbargirrar, *which concludes expressed numbers;
anything beyond one hundred is* larbargirrar larbargirrar, *signifying a crowd
beyond counting, and is always accompanied by repeated opening and shutting
the hands (Dawson [1881], p. xcviii).*

Thus there is no verbal indication in the words themselves of the possibility of going on indefinitely.   On the other hand, the very act of opening and shutting the hands appears, simultaneously, to recognize that possibility.

It therefore seems reasonable to interpret the above as showing that it was only gradually that the idea of unending repetition came to be accepted.

Another example of the gradual admission of indefinitely extended counting and perhaps even the idea of infinity is quoted by Seidenberg.

*The number of terms that the Pukapukan uses to indicate high numbers is
interesting though it is a little hard to see the function of high numerical
concepts in an atoll culture.*

*It was a favorite jest among informants that the Pukapukan could count
to a higher power than we could; proof of this they argued was not only the
presence of words indicating progression to infinity, but also the ability
of the culture hero Maui to find Pukapukan words which enabled him to
enumerate the stars in the sky, the fish in the sea, the sands on the beach,
and so forth.[67]*

This problem of indefinite repetition now begins to have two facets. One is the idea of infinity, the other the idea of placing one thing (be it a tally as in tally counting or a man as in Eskimo counting or whatever) after another.   For the latter what seems to be crucial is the presence of order.[68] By this we do not necessarily mean any abstract idea of order; it is sufficient to have the process of arrangement.   If acts of sequential arrangement are used the possibility of an arbitrarily long sequence is assured.   This was always a possibility, even in the 2-counting languages where one counts one, two, two-one, two-two, two-two-one, etc.. Even so, the move to counting indefinitely far is still a gradual move and not an abrupt one, as Lévy-Bruhl appears to have thought.   As Cassirer puts it:

*We find that analogously to what has been observed in the process of
counting, language does not abruptly juxtapose an abstract category of unity
to an abstract of plurality, but finds all manner of gradations and*

transitions between them. The first pluralities that it distinguishes are not general but specific pluralities, with a distinctive qualitative character. Aside from the use of dual and trial, many languages employ a double plural; that is, a narrow form for two or a few objects, and another for many objects. This usage, which Dobrizhoffer found in the language of the Abipones[69], has its exact counterpart in the Semitic languages, for example the Arabic.[70] In his account of the plural forms in Arabic (which beside the dual has a limited plural for 3 to 9 and a multiple plural for 10 and over, or for an indeterminate number of objects), Humboldt remarks that the underlying conception, which in a sense situates the generic concept outside the category of number, so that both singular and plural are distinguished from it by inflection, must "undeniably be called a very philosophical one".[71] In truth, however, this generic concept does not seem to be conceived in its determinate generic character and thus raised above the category of number; on the contrary, the category of number does not yet seem to have entered into this form. The distinction which language expresses by singular and plural has not been taken up into the genus; indeed, it has not yet been sharply drawn; the quantitative opposition of unity and multiplicity has not yet been overcome by a qualitative unity which encompasses them both, because for the present this opposition has not yet been clearly determined (Cassirer [1953], p.237).

To put it in somewhat less Germanic language: Cassirer appears to regard the distinction between one and many, between unity and plurality, as more fundamental than that between one and any particular number (greater than one). In the example of the Aborigines quoted from Dawson (see above), the distinction is (roughly) between numbers less than a hundred and those above. In any case, the idea of *indefinite* or endless repetition is mediated by a stage which involves a significant amount, but not an unending amount, of repetition.

This difference, which we may express as the difference between being able to go on as far as you like and being able to go on forever, in historic times first appears, so far as we know, in Aristotle to whom we shall shortly turn. But the idea that once one has repeated a counting move, one can repeat it again, and again, and again, and ... is the central idea which led to the nineteenth-century formalization of the natural numbers by Dedekind.

In the next chapter we trace the two-thousand-year development of this idea from the time of Aristotle to that of Dedekind.

Chapter II

Historic Times

1. The Greeks

In the first chapter we saw how gradually the natural numbers developed. There the idea of unending repetition, used but not defined, was clearly crucial to the development of a system which could include arbitrarily large numbers. The formal definition appeared only very slowly and in this chapter we shall see it gradually emerging. We start with the Greeks, where there is no definition but there are constructions which we today would formalize inductively (or recursively, see below). From then on there is a gradual development.

The Pythagoreans held the view that numbers[1] were constituents of everything (Aristotle, *Metaphysics*, 1080a) yet we find Aristotle, in his *Physics* III, 4.203a, describing the way the Pythagoreans constructed numbers using gnomons. That is to say, numbers were considered as diagrams of dots or pebbles or alphas thus :

$$
\begin{array}{ccc}
\alpha & \alpha & \alpha \\
\alpha & \alpha & \alpha \\
\alpha & \alpha & \alpha
\end{array}
$$

for the odd numbers and

$$
\begin{array}{cccc}
\alpha & \alpha & \alpha & \alpha \\
\alpha & \alpha & \alpha & \alpha \\
\alpha & \alpha & \alpha & \alpha
\end{array}
$$

for the even numbers, where succeeding odd (even) numbers are those that fill the gnomon or L-shaped annexe.

That such constructions were important in Greek mathematics seems clear from a number of references Aristotle makes. For example, in his *De Caelo* 279b33, while discussing arguments about the generation of the universe, he refers to arguments of Xenocrates and the Platonists[2] thus :

They say that in their statements about its generation they are doing what geometricians do when they construct their figures.

That is, they are just continuing the same process.   In the same way in the *Categoriae* Aristotle writes in translation :

> The square, for instance, if a gnomon is applied to it, undergoes increase but not alteration[3], and so it is with all other figures of this sort.   Alteration and increase, therefore, are distinct (Categories, 15a 30).

That is, the shape remains constant while the size increases.   This indicates some idea of uniformity-of-process.

It appears that, in Greece, the idea of numbers being generated from one by a splitting process first occurs with Plato, for Aristotle reports :

> ... the units in the ideal 2 are generated at the same time, whether, as the first holder of the theory[4] said, from unequals (coming into being when these were equalized) or in some other way (Metaphysics, 1081a).

On the other hand Aristotle holds the view that number is prior to the dyad.   He does not accept that the basis of number is the One together with the indefinite dyad as Plato did (cf. *Metaphysics*, 1080b).

> And in general the arguments for the [Platonic] Forms destroy things for whose existence the believers in Forms are more zealous than for the existence of the Ideas ; for it follows that not the dyad but number is first, and that prior to number is the relative, and that this is prior to the absolute[5] ... (Metaphysics, 1079a).

We have already echoed Aristotle's conclusions about numbers in the previous chapter, §8, (though his remarks above seem neither fully justified nor justifiable), for he writes at the very end of his *Metaphysics* (1093b) :

> The fact that our opponents have much trouble with the generation of numbers and can in no way make a system of them seems to indicate that the objects of mathematics are not separable from sensible things as some say, and that they are not the first principles.

Aristotle was also keenly aware of the differences between the various senses of unending, or indefinite, continuation.   Indeed, his distinction between potential and actual infinities is still a matter of fundamental concern for philosophers of mathematics.[6]

Aristotle writes in his *Physics* (204a) :

> We must begin by distinguishing the various senses in which the term 'infinite' is used.
> (1) What is incapable of being gone through, because it is not its nature to be gone through (the sense in which the voice is 'invisible').
> (2) What admits of being gone through, the process however having no termination, or

(3) What scarcely admits of being gone through.
(4) What naturally admits of being gone through, but is not actually gone through or does not actually reach an end.

It seems extremely likely that he was quite aware that in counting one needs not a completed infinity of numbers but only the facility to count as far as one might wish. Some sort of evidence for this is to be found slightly later in Aristotle, though it must be pointed out that the context is geometric rather than numerical and that the Greeks were keenly aware of a difference between length and number.

In point of fact they do not need the infinite and do not use it. They postulate only that the infinite straight line may be produced as far as they wish (Metaphysics, 206b).

Now Aristotle lived in the fourth century B.C., but this attitude persists in Euclid, who lived at the beginning of the third century B.C.. In the latter's works we do not find Euclid speaking of, for example, "parallel lines meeting at infinity" or of there being "an infinite number of prime numbers". Instead we find the comment that parallel lines do not meet, however far produced[7], and that

Prime numbers are more than any assigned multitude of prime numbers (Proposition 20, Book IX, [1925], vol.2, p.412).

That is to say, given any collection of prime numbers we can find a larger prime number. We include Euclid's proof of this proposition now for we shall wish to refer to it later. We note that Euclid represents an arbitrary collection of primes, by writing   A, B, C   whereas we would write A, B, ..., C. He appears to indicate that the reader could easily supply the proof for any other given (particular, finite) collection of prime numbers. Euclid's presentation is as general as was reasonable or possible at that time, for although the Greeks invented punctuation, the use of the ellipsis "..." came much later. Euclid writes :

Let   A, B, C   be the assigned prime numbers, I say that there are more prime numbers than   A, B, C.

For let the least number	A —
measured by A, B, C be taken,	B — G ——
and let it be DE ;	C ——
let the unit DF be added to DE.	E ——————— D — F

Then EF is either prime or not.
First, let it be prime ; then the prime numbers   A, B, C, EF   have been found which are more than   A, B, C.   Next, let EF not be prime; therefore it is measured by some prime number.   [VII. 31]   Let it be measured by the prime number G.   I say that G is not the same with any of the numbers A, B, C.   For, if possible, let it be so.   Now A, B, C measure DE ; therefore G also will measure DE.   But it also measures EF.   Therefore G, being a number, will measure the remainder, the unit   DF : which is absurd.   Therefore G is not the same with any one of the numbers A, B, C.   And by hypothesis it is prime.   Therefore the prime numbers   A, B, C, G   have been found which are more than the assigned multitude of   A, B, C.
Q.E.D. (ibid., vol.2, p.412).

This proof is what Freudenthal ([1953], p.22, 30) calls "quasi-general" (*quasi-allgemein*) in his history of induction[8]. In fact, the present-day proof proceeds :

Suppose p is the largest of a given (finite) set of primes. Consider p!+1, then it has no divisors ≤ p except 1, since all other numbers from 1 up to and including p leave remainder 1. Hence the smallest prime dividing it is greater than p (Euclid's result).

This proof conceals (or contains) an inductive premise : Suppose p is the n-th prime (in some list). The proof then shows there exists a bigger prime, thus extending the list to an (n+1)-st prime. Hence by induction there are infinitely many primes.

What Euclid in fact claims is that beyond any given (finite) collection of primes there is a greater prime. (Euclid also uses this approach elsewhere in his work, e.g. Book IX, Proposition 36.) Thus the concept involved here is that of finding a larger number (satisfying certain conditions) than any previously obtained. We are therefore in the realm of Aristotle's second sense of "infinite".[9] The idea of a completed infinite sequence or series does not appear. What is present, however, is the notion of an arbitrarily long finite sequence. Thus in Euclid's Book IX, proposition 8, we read : "If as many numbers as we please ... be in continued proportion...".

What these examples seem to show is simply a great awareness on the part of Euclid of the nature of the natural numbers even though he does not have the formal apparatus of induction.[10] Since Euclid's proofs do not proceed from a formulation of induction, it is not particularly surprising that he should not have included any such principle in his basic machinery.

We find Euclid using something akin to our notion of mathematical induction with no formal notice of this point. After Euclid we shall turn to Nicomachus of Gerasa who, in about the second century A.D., wrote an influential and very long-lasting book on arithmetic which was freely translated into Latin by Boethius (*c.* 480-524/5). In their introduction to D'Ooge's translation of Nicomachus, Robbins and Karpinski write :

It has long been recognized that the De Institutione Arithmetica of Boethius ... has so little claim to originality that it may be called a translation of the Introduction of Nicomachus.[11] The judgment is a just one ([1926], p.132).

They go on to say that Boethius did as he had promised "expanding here and contracting there".

Nevertheless, Boethius' version "became and remained the source through which the Latin-speaking portion of the world knew Nicomachus" (*ibid.*,p.125) (though the Arabs learned of Nicomachus through a partial translation by Thabit ibn Qorah (836-901 A.D.) *ibid.*).

2. Nicomachus

Book 1 of Nicomachus' *Introduction to Arithmetic* is mainly concerned with the analysis of numbers (into types, e.g. even and odd). In the second book we find a more synthetic approach. At the beginning of this book we find :

We wish also to prove that equality is the elementary principle[12] *of relative number ; for of absolute number, number per se, unity and the dyad*[13] *are the most primitive elements, the least things out of which it is constructed, even to infinity, by which it has its growth, and with which its analysis into smaller terms comes to an end (ibid., p.230).*

Now Nicomachus used the Greek numeral system employing letters (sometimes with diacritics) of the old Greek alphabet[14] (α = 1, β = 2,..., ι = 10, κ = 20,..., ρ = 100,..., ω = 800, λ = 900). Having mentioned these, Nicomachus continues (in translation) :

On the other hand, the natural, unartificial, and therefore simplest indication of numbers would be the setting forth one beside the other of the units contained in each. For example, the writing of one unit by means of one alpha will be the sign for 1 ; two units side by side, that is, a series of two alphas, will be the sign for 2 ; when three are put in a line it will be the character for 3, four in a line for 4, five in a line for 5, and so on (Nicomachus [1926], p.237).

That is, 1 is represented by α, 2 by $\alpha\alpha$, 3 by $\alpha\alpha\alpha$, etc.. This is the start of Nicomachus' construction of figured numbers, that is to say, the representation of numbers by geometric arrangements.

Take, for example, 1, 4, 9, 16, 25, 36, 49, 64, 81, 100 ; for the representations of these numbers are equilateral, square figures, as here shown ; and it will be similar as far as you wish to go :

| 1 | 4 | 9 | 16 | 25 |

It is true of these numbers, as it was also of the preceding, that the advance in their sides progresses with the natural [number] series (ibid., ch. IX, p.242).[15]

But apart from this regular progression, he also speaks of the basic generation of number arising from "otherness", that is, the idea of a starting point plus splitting. Again we quote the translation of Nicomachus :

For the ancients of the school of Pythagoras and his successors saw 'the other'[16] and 'otherness' primarily in 2, and 'the same' and 'sameness' in 1, as the two beginnings of all things, and these two[17] are found to differ from each other only by 1. Thus 'the other' is fundamentally 'other' by 1, and by no other number, and for this reason customarily 'other'[18] is used, among those who speak correctly, of two things and not of more than two (ibid., p.254-55).

And again :

In this way, then, all numbers and the objects in the universe which have been created with reference to them are divided and classified and are seen to be opposite one to another, and well do the ancients at the very beginning of their account of Nature make the first subdivision in their cosmogony on this principle. Thus Plato[19] mentions the distinction between the natures of 'the same' and 'the other', and again, that between the essence which is indivisible and always the same and the one which is divided ; and Philolaus[20] says that existent things must all be either limitless or limited, or limited and limitless at the same time, by which it is generally agreed that he means that the universe is made up out of limited and limitless things at the same time, obviously after the image of number, for all number is composed of unity and the dyad, even and odd, and these in truth display equality and inequality, sameness and otherness, the bounded and the boundless, the defined and the undefined ([1926], p.258-259).

Thus Nicomachus clearly regards 1 as the starting point for numbers and then follows the Pythagorean and Platonic view (see above, §1) and regards 2 as signifying the division or partition into parts which are "other". He clearly brings out the separation of two fundamental aspects of the natural number series which are still current : the beginning (with 1) and the augmentation step (which is repeated arbitrarily often.[21]

In contrast to his treatment of square numbers (see above, §2), his treatment of triangular numbers exemplifies these two fundamental aspects. He describes the genesis of triangular numbers in Chapter VIII of Book II. Triangular numbers are those obtained by making triangles of, say, alphas, thus :

and thus are the numbers 1, 3, 6, 10, 15, ... ([1926], p.241).

The triangular number is produced from the natural series of number set forth in a line, and by the continued addition of successive terms, one by one, from the beginning ; for by the successive combinations and additions of another term to the sum, the triangular numbers in regular order are completed (ibid., p.241).[22]

The rest of Nicomachus' book is concerned with specific types of number beyond triangular and square numbers and is not our concern here.

Robbins and Karpinski in their introduction to D'Ooge's translation write :

> *Judged by the standards of the mathematician, Nicomachus cannot rank with the leaders of the science even as it was known in antiquity ; estimated, however, by the number of his translators, scholiasts, commentators, and imitators, he is undoubtedly one of the most influential. From his own day until the sixteenth century, among the Greeks, the Latins, and the Arabs, there was scarcely a place where he was not honored as an arithmetician, or a time when learned men failed to regard his work as the basis of the science (ibid., p.124).*

Nicomachus is one of the very few authors to have been handed down directly while other authors either survived in Arabic translations or were simply mislaid before being rediscovered hundreds of years later.   Euclid suffered such a fate : Nicomachus did not.   So we can trace his influence and a continuity of idea right up to the Renaissance.   Even his ideas, limited and derivative though they usually were, were better than those of some of his commentators.

3. The Dark Ages

Nicomachus, as we noted above, was freely translated into Latin by Boethius, thus making Nicomachus' work available to all intellectuals in Western Europe.   The work was known to the encyclopaedist Isidore[23] of Seville (d. 636).   Now Isidore appears somewhat muddled at times and his passage on the natural numbers shows this.   He basically seems to fail to distinguish between the *collection* of all numbers being infinite (see the end of the passage below) and a number (necessarily not a natural number) being infinite.   In his *Etymologiae* we read :

How many infinite numbers exist[24]

> *But it is most certain that numbers are infinite, since at whatever number you may think of making an end, I say not only that that same can be augmented by adding one, but however large it be, and however huge a multitude it contains, by the very idea and science of numbers it can not only be duplicated, but even multiplied.   So truly a number is limited each to its properties, so that none of them can be equal to any other.   And therefore they are unequal and distinct amongst themselves, and every one on its own is finite, and all are infinite ([1911], Lib. III, ix).[25]*

Thus Isidore recognizes that one can always produce a larger natural number.   It appears also that he believes the collection of natural numbers to be infinite.

Isidore has nothing more to say about *all* numbers though he does discuss types of number in the style of Nicomachus.   By the thirteenth century we find Jordanus Nemorarius being much clearer – and closer to Euclid.

Euclid (in Book VII) had given definitions of units and parts (that is, submultiples) and Jordanus Nemorarius in his *Arithmetic* [1496] proceeds in similar fashion, except for his third definition. Nemorarius writes :

A unit is the discrete existence of a thing by itself. A number is a collective quantity of discrete things. A series of numbers is said to be natural if its computation of its members is done by the addition of one[26,27,28] *([1496], f. a2r).*

So here we have an explicit presentation of a "natural series" of numbers. The rest of Nemorarius' book is concerned with Pythagorean-style number theory and with proportions. He approaches numbers as objects with special properties and he enunciates those properties. One has the impression that these properties are both unknown and surprising. The limited degree of sophistication in the work may be judged by the following. One of the most advanced, but unproved, theorems is that, of the three means of two numbers, the arithmetic mean is the maximum, the harmonic the minimum and the geometric between the other two. So it is perhaps best to say that Nemorarius (or his predecessors) *discovered* that natural numbers are obtained by the successive addition of one. We shall later find faint echoes of this in Dedekind (see below, §6), where he singles out the successive application of a process (function), to obtain all natural numbers, as being one of the essential characteristics of the natural numbers.

When we turn to Jordanus' contemporary, Fibonacci (otherwise known as Leonardo Pisano, *c.* 1179-post 1240), we find much more sophisticated mathematics. Although, unlike Jordanus, Fibonacci does not use letters for arbitrary numbers, he does prove general results. For example, he uses two approaches to proving what in present-day notation we should write as the identity $n^2 + (2n+1) = (n+1)^2$. In some of his work the proofs appear to be clearly inductive, in others geometric or algebraic. Thus in discussing the series of square numbers, Fibonacci writes :

I thought about the origin of all square numbers and discovered that they arose from the regular ascent [or increase] of odd numbers. For unity is a square and from it is produced the first square, namely 1 ; adding 3 to this makes the second square, namely 4, whose root is 2 ; if to this addition [that is, sum] is added a third odd number, namely 5, the third square will be produced, namely 9, whose root is 3 ; and so [it is that] the sequence and series of square numbers always takes rise through the regular addition of odd numbers [1 + 3 + 5 = 9, 1 + 3 + 5 + 7 = 16,...].

Thus when we wish to find two square numbers whose addition produces a square number, I take any odd square number as one of the two square numbers and I find the other square number by the addition (collectione) of all the odd numbers from unity up to [but excluding] the odd square number [already selected]. For example, I take 9 as one of the two squares mentioned ; the remaining square will be obtained by the addition of all the odd numbers below 9, namely 1, 3, 5 and 7, whose sum is 16, a square number, which when added to 9 gives 25, a square number [9 + (1 + 3 + 5 + 7) = 9 + 16 = 25 = 5²] (Liber quadratorum [1768], vol.ii, p.253).[29]

In this short book of indeterminate analysis Fibonacci uses the above technique frequently. In the first part of our quotation he is giving an inductive procedure - inductive in the Aristotelian and philosophers' sense. He is showing how the series of odd numbers and the series of squares progress in tandem. He leaves it to the reader to prolong the series as far as the latter wishes. Thus the following diagram is perhaps a good way to translate Fibonacci.

Squares   1   4   9   16
               + = + = + = ...
Odd numbers   3   5   7

It is certainly clear from an intuitive point of view that taking an odd square, then adding all the odd numbers below it, gives another square. (Another way is to observe that the gnomons we discussed in §1 of this chapter contain the successive odd numbers 1, 3, 5,... and that taken together successively they build the square numbers 1, 4, 9,..., cf. the first diagram in this chapter.) The general argument, in modern notation, makes it look rather different:

Suppose m is an odd square, say $m = (2n+1)^2$. Then $m = 2r+1$ for some r. The sum of the odd numbers < m is

$$1 + 3 + \ldots + (2r-1) = r^2$$

(as we know from the previous discussion). Hence $r^2 + m = r^2 + (2r + 1) = (r + 1)^2$ which is again a square.

Neither this sort of argument nor anything like it was used by Fibonacci, though he was quite capable of extended arguments (see his treatment of the solution of a cubic, below, Chapter III, §8). It seems much more likely that Fibonacci simply experimented with number sequences and noted certain regularities (c. the approach of Wallis discussed below, §5).

Later ([1857,1862], p.225) Fibonacci gives a geometric demonstration of how the series of squares is obtained by adding the odd numbers "beginning from one, going on to infinity".[30]

Scientists use the word "induction" to mean that one is basing one's claims on the results of previous observations or experiments. We shall see how this use of "induction" was gradually transformed into the formal method of proof by induction we know today. Somewhat similar arguments may be found among the Arab writers. Indeed, it seems that it was through his stay in North Africa and his travels around the Mediterranean that Fibonacci learned most of his mathematics. As an example of the Arab work we quote, abbreviating slightly, from the Arabic algebra al-Fakhrî (c. 1010 A.D., [1853], p.61), translated into French by Woepcke.

Theorem.

$$1^3 + 2^3 + \ldots + 10^3 = (1+2+3+\ldots+10)^2$$

Demonstration.

$$(1+2+3+\ldots+10) = 55 = 45 + 10,$$

$$(45+10)^2 = 45^2 + 2.10.45 + 10^2$$

$$= 45^2 + 10^3,$$

$$45^2 = \ldots = 36^2 + 9^3$$

$$36^2 = \ldots = 28^2 + 8^3.$$

Continuing thus one verifies the theorem.[31]

It is curious that the author of al-Fakhri proceeds step by step downwards from 10 but he does make the step-by-step process clear. In the move to present-day formulations of mathematical induction there was a gradual shift to the two-part process of base case plus induction step.

4. Renaissance

In a study of mathematical induction Freudenthal [1953] has distinguished various stages. One of these is the "quasi-general" ("quasi-allgemein") stage where a proof is presented using specific numbers but where it is clear that the proof works for arbitrary numbers. Obviously the use of letters to stand for arbitrary numbers makes this clearer.

It is generally accepted that Jordanus Nemorarius, who lived in the thirteenth century, was the first to use letters systematically for numbers.[32] Letters were also used by Rabbi Levi ben Gershon (otherwise known as Leo de Bagnolas, Maestro Leon or Gersonides) in his Hebrew work *Maasei Hoshev* (The Work of the Calculator) of 1321.[33]

Gersonides gave a proof of the theorem from al-Fakhri quoted above. In the proof of this theorem, which is proposition 42, he only gives a quasi-general proof for $1^3 + 2^3 + 3^3 + 4^3 + 5^3 = (1 + 2 + 3 + 4 + 5)^2$. However, the lemma that he needs is clearly quite general.

Proposition 41. The square of the sum of the series of integers beginning from one up to a given integer is equal to the cube of the given integer plus the square of the sum of the series of integers from one up to the predecessor of the given integer (Rabinovitch [1969], p.244).

In proving the lemma he uses the series of numbers a, b, c, d, e but this can clearly be interpreted, in modern notation, as a, b, c, d,..., e and the proof is completely general (cf. §3 above).

A clearer example is in his sequence of theorems establishing that the product of n numbers is independent of their order and the order of multiplying : a sort of combined version of the generalized associative and commutative laws (Rabinovitch [1969], p.240-241). In outline he proceeds as follows :

Proposition 9. If one number multiply another which is itself the product of two given numbers the result is the same as when the product of any two of these three factors is multiplied by the third.

$$[a(bc) = b(ac) = c(ab)]$$

Proposition 10. If one number multiply another which is itself the product of three given numbers the result is the same as when any one of these numbers multiplies the product of the other three.

$$[a(bcd) = b(acd) = b(dca) = ...]$$

Thus proposition 10 is essentially proposition 9 with "two" replaced by "three". After proving proposition 10, Gersonides states :

In this manner of rising step by step [the Hebrew word is Hadragah], it is proved to infinity. ... And because of this, the result of multiplying one number by a product of other numbers contains any one of these numbers as many times as the product of all the others.

$$[a(bc...z) = b(ac...z) = etc.]$$

Rabinovitch sums up Gersonides's work thus :

It seems fair to conclude that R. Levi Ben Gershon found the method of mathematical induction used sporadically by various authors [some presumably Arab] in different contexts. However, he recognized its special usefulness and formalized the structure of the proof, giving it an apt and descriptive name and applying it to obtain many different results ... (Rabinovitch [1969], p.248).

The last sentence quoted seems to state the case with a precision inappropriate to the fourteenth century, but it is clear that Gersonides was moving towards formalization and a precise description or definition.

Rabinovitch's work was, at least in part, a response to that of Freudenthal [1953] who, in turn, had taken to task Vacca, who had claimed that Maurolico (1494-1575) was "the First Discoverer of the Principle of Mathematical Induction" - to quote part of the title of one of Vacca's papers [1909].

Maurolico lived about two hundred years after Gersonides, but their work has some items in common, including the theorem from al-Fakhrī. One suspects that many of these items ultimately derived from the Arabs. On inspection it appears that Maurolico is not as precise as Gersonides.

Maurolico's *Arithmeticorum Libri Duo* [1575][34] begins with a table of numbers of various kinds (triangular, square, etc.) and then follows :

Formation of the numbers in the preceding table

Roots are formed from unity by the continual addition of units (ibid., f. a.).[35]

Thus echoing Nemorarius. The rest of the table is explained in the same way.

In his propositions he gradually builds up an analysis of the numbers in the table and his thirteenth proposition reads :

Each square [number] added to the next odd [number] makes the next square [number]. [I.e. $n^2 + (2n+1) = (n+1)^2$] (ibid., p.7).[36]

Freudenthal ([1953], p.24) describes Maurolico's proof as "quasi-general", that is, in our terminology, a proof of the general form rather than an inductive one. What Maurolico does is simply to give the example :

The fourth square is 16. When conjoined with the odd [number] in the fifth place, i.e. 9, makes the fifth square [i.e. 25] (ibid., p.7).[37]

This case he establishes using his earlier propositions (the sixth and twelfth). Now in his fifteenth proposition Maurolico writes :

By the aggregation of the odd numbers taken in proper order beginning with unity the series of square numbers can be constructed collateral with those odd numbers.[38]

Maurolico here uses his thirteenth proposition, writing :

For by the previous prop[osition] 13, unity first of all with the next odd [number] makes the next square, that is, 4. And this 4. the second square, with the third odd [number] that is, 5. makes the third square, that is, 9. And likewise 9, the third square, with the fourth odd [number], that is, 7, makes the fourth square, that is, 16 and so successively to infinity, the 13th [proposition] always being repeated, so that what we are now proposing is demonstrated.[39]

Freudenthal argues that this is indeed an "induction inference".[40] He argues that a proof is "quasi-general" if the result required is obtained by treating an arbitrary number as a variable. Thus a proof of $n^2 + (2n+1) = (n+1)^2$ would normally nowadays be "quasi-general" (using the identity $(n+1)^2 = n^2 + 2n + 1$). A proof that $\sum_1^n (2m-1) = n^2$ would most likely be "inductive" ("induktiv") in the form

$$\sum_1^{n+1}(2m-1) = \left[\sum_1^{n}(2m-1)\right] + \left[2(n+1)-1\right]$$

$$= n^2 + (2n+1) = (n+1)^2.$$

Maurolico's *proof* certainly is closer to the latter, but in the margin he puts

$$\begin{array}{l} 1 \\ 3 \end{array} \} \, 4 \, \Big\} \, 9 \\ \begin{array}{l} 5 \\ 7 \\ 9 \end{array} \Big\} \, 16 \, \Big\} \, 25$$

thereby indicating his view.   In fact it is rarely that his marginal picture indicates anything like an induction.,   Thus for his proof of the next proposition (the fourteenth) which is, in our notation, $(n+1)^2 = n^2 + 2n + 1$, he puts

$$\begin{array}{l} 4 \\ 4 \end{array} \} \begin{array}{l} 16 \\ 8 \\ 1 \end{array} \Big\} \, 25$$

The progressive picture only occurs once more in the thirty or so pages of Part I of Book I (for proposition 51 [1575], p.22).   Progressive pictures occur infrequently in the rest of the work too, though there are lots of tables listing first, second, third, etc. numbers of various kinds.

Now Maurolico does not progress from n to n + 1 ; he simply argues the first case and then argues "and so successively to infinity" each time, that is, for each successive number, using a lemma again (his proposition 13).   This phrase echoes Gersonides's "rising step by step".   However, in the second book he does give general proofs for some theorems.   In the preface to his second book, he says that he will give general procedures and proofs (*ibid.*, p.83-84) and he proceeds to deal with numbers, lines, surfaces, etc. simultaneously.

He does in fact give a general proof of $(a+b)^3 = a^2 + b^3 + 3a^2b + 3b^2a.$

Let ab. be a quantity divided into two parts, say a. and b. I say that the cube of the whole ab. is equal to this : that is, the cube of that a. and the cube of that b. and the triple of the product of the square of a. multiplied by b. together with the triple of the product of the square of b. multiplied by a.[41]

He proves this using Euclid II.4.   Thus Maurolico had the machinery for proving general propositions in literal[42] algebra such as $(n+1)^2 = n^2 + (2n+1)$.   We have no evidence that he used it.

In the light of all this, it remains debatable whether we identify Maurolico's progressive proofs or Gersonides's "rising step by step" with induction in the form : P(1) and whenever P(n) holds, then P(n+1) holds, hence, for all numbers n, P(n). The problem here is two-pronged. First, until a *definition* of induction is given, there is no chance of deciding whether a particular inference is an induction or not. The second is that, as we hope we are showing, the development of the idea and its subsequent formulation (or formalization) have been remarkably gradual. One thing, however, is clear: with the very process of proceeding from one number to the next (in a sequence), as both Gersonides and Maurolico do, one is coming closer to induction than the bare idea of the natural numbers being formed by successive addition of one, as Jordanus did (see above, §3).

In his further remarks Freudenthal[43] goes on to trace possible inductions back to Euclid, but we shall move on to the seventeenth century having, we trust, put into perspective Freudenthal's claim that Vacca had overstated his case in attributing the *origin* of mathematical induction to Maurolico[44] and also having seen that Gersonides has a stronger case. Note in particular that Maurolico certainly did not use the name "mathematical induction"[45] or, indeed, any special name[46], while it is debatable whether Gersonides regarded "rising step by step" as a name or simply an informal guide.

5. The Seventeenth Century

In the following century, the seventeenth, we *do* find quite explicit use of the word "induction", though not always in the neat and precise forms given by Pascal and Bernoulli (see below). Wallis discusses the value of

$$\frac{0 + 1 + 4 + 9 + \ldots + n^2}{n^2 + n^2 + n^2 + n^2 + \ldots + n^2}$$

in his *Arithmetica Infinitorum* ([1699], p.336). For successive values of n he gets smaller and smaller values and concludes that eventually the value vanishes (*evaniturus fit*), i.e. becomes zero. However, the sense in which Wallis uses "induction" is that of scientists rather than of mathematicians. That is to say, he relies on repeated "experiments" giving the same result rather than a mathematical (viz. logical) proof of the result. Thus in his *Algebra* he writes :

Those Propositions in my Arithmetick of Infinites, are (some of them) demonstrated by way of Induction : Which is plain, obvious, and easy ; and where things proceed in a clear regular Order, (as here they do,) very satisfactory, (to any who hath not a mind to cavil;) and shews the true natural investigation. Which to me, is much more grateful and agreeable, than the Operose Apagogical Demonstrations, (by reducing to Absurdities or Impossibilities,) which some seem to affect ; and which was much in use amongst the Ancients, for reasons which now (in great measure) are ceased since the introducing the Numeral Figures, and (much more) since the way of Specious Arithmetick ([1699], p.298).

Thus Wallis is not providing a formal proof at all!   This example shows that "induction", in the sense of "mathematical induction", was not familiar to Wallis.

We next turn to Pascal, who meets the most stringent requirements when one is seeking the first to recognize, define and formally use the principle of mathematical induction.   This is exemplified in Pascal's triangle.

Now Pascal's triangle was discovered before 1000 A.D. in China.   Jia Xian is usually credited with the discovery and an extensive mathematical account was given by Zhu Shijie around 1300 A.D. (see Li Yan and Du Shiran [1987], p. 117f.).   The Chinese, however, were more concerned with algorithms than with proofs and there is no suggestion in Zhu Shijie's work of a method of proof; he is only interested in a method for solving polynomial equations of the n-th degree.

Pascal, however, in his *Traité du triangle arithmétique* [1665] of 1654 not only uses an inductive proof, he even draws attention to it as a *method* of proof.   In his proof of his *Conséquence Douzième* he says :

Although this proposition has an infinite number of cases, I shall give a very short demonstration of it based on 2 lemmata.   The first, which is self-evident, is that this proportion holds in the second base [i.e. hypotenuse of the second smallest triangle]; for it is quite clear that ϕ is to σ as 1 to 1.

The second is that if this proportion holds in some arbitrary base then it necessarily also holds in the following base.
Whence we see that it is necessary in all the bases : for it is in the second base, by the first lemma, therefore by the second it is in the third base, therefore in the fourth, and so on to infinity (ibid., p.7).[47,48]

Here the "bases" are the hypotenuses of the familiar Pascal triangles written in the form

$$
\begin{array}{cccccc}
1 & 1 & 1 & 1 & 1 & 1 \\
1 & 2 & 3 & 4 & 5 & \\
1 & 3 & 6 & 10 & & \\
1 & 4 & 10 & & & \\
1 & 5 & & & & \\
1 & & & & &
\end{array}
$$

and the induction proceeds from one *base* (hypotenuse) to the next, so the bases may be labelled first, second, etc..

Thus here we do have a quite explicit form of induction in the sense that both stages, the initial case and the progression from n to n+1, are both produced and then the inference made that the result holds for all natural numbers. Of course, Pascal does not *call* this process "mathematical induction".

In fact, for earlier propositions about his triangle Pascal proves his statements by informal induction in the same style as Wallis. Thus he establishes (*consequence huictiesme* [1665], p.5) that (in our parlance) the sum of the binomial coefficients for exponent n is 2^n. He proceeds :

For the first base is unity. The second is double the first, therefore it is 2. The third is double the second, therefore it is 4. And so on to infinity.[49]

He then goes on with a note which is in fact a generalization of the above.

When he comes (on p. 7) to the proof of what we write as[50]

$$C_{n\ m-1} : C_{n\ m} = m-1 : n-m$$

he gives an example and then proceeds formally.

Although this proposition has an infinite number of cases, I shall give a very short demonstration of it based on 2 lemmata ... [and continues as above].[51]

He *then returns* to the Wallis-type exposition and does not produce the 2-lemmata argument again. He does use it in the third tract, "*Usage du Triangle Arithmétique pour les partys ...*" (Pascal [1665], p.8, for a different proposition.

About the same time, Fermat used a method of proof which we now know is equivalent to induction. I do not know if such an equivalence was known or meaningful to Fermat. His method was the method of infinite descent ("descente infinie ou indéfinie"). The method is : Assume there is a counterexample to an assertion. Show that there is a smaller counterexample. This gives a contradiction since any decreasing sequence of natural numbers must eventually stop. The contradiction thus produced establishes the theorem.[52]

With Jacob Bernoulli, however, we again find the principle of induction clearly enunciated. The *Acta Eruditorum* of 1686 includes an excerpt from a letter in which he writes, after referring to induction in Wallis's *Arithmetica Infinitorum*:

E.g. Let us investigate whether the ratio of the series of numbers succeeding each other in natural progression and starting from zero to the series of just as many [numbers] equal to the maximum is the subduple [i.e. (0 + 1 + ... + n) : (n + ... + n) = 1 : 2]. Let the entity have been analyzed up to some indefinite point; the last term, on which I have paused in order to analyze it, I call a: and the number of terms from the initial zero will be greater by one, namely a + 1: and so the sum of all those equal to the last will be aa + a, to which when it is assumed by induction that the sum of the progression will have born the subduple [ratio], it [the sum of the progression] will be $\frac{aa + a}{2}$ *. Now increase the series by one term more, and the added term will be* $aa + 1$, *which added to the sum of the preceding [terms]* $\frac{aa + a}{2}$ *produces* $\frac{aa + 3a + 2}{2}$ *[as] the sum of the whole progression: but since the number of terms is now a + 2, the sum of terms all equal to the last added term will be aa + 3a + 2 which in like manner bears the duple [ratio] to the sum of the progression. But if now that term which was lately called a + 1 is called a, and is added new to the progression which will be a + 1, the same demonstration will be valid: when therefore [for a series] bearing the subduple ratio to some arbitrary series it can be inferred then the same holds for a series augmented by one term, and even to a series augmented by two, three, etc. infinitely many terms, it always follows, but if this property can be carried over by induction to a few series, likewise it is common to all (Bernoulli [1686], p.360).*[53]

But having given this specific example, he goes on to note that the *method* is quite general.   The different formulation using a letter (*a*) for an arbitrary number makes this a clear use of induction in the modern sense. Due to Viète (1540-1603), literal algebra had developed richly, though it was, as we have noted above, used a little by Maurolico in 1557 and even earlier, though to a very limited extent, by Jordanus Nemorarius.

Thus by 1686 the idea of mathematical induction is both explicit and used.   In the succeeding century, though not very widely used as a name, induction was employed though so, too, was Wallis's "induction" (scientific or experimental induction as we say).[54]   It was not, until the latter half of the nineteenth century that it started to play a central rôle.   The name "mathematical induction" appears to be due to De Morgan in 1838 (see Cajori [1918]).

In the mid-nineteenth century another element enters our discussion: the notion of one-one correspondence.   However, before turning to this notion we shall consider Dedekind's refinement of induction.

6. Dedekind and Peano

The development of analysis led, in the mid-nineteenth century, to a questioning of the foundations of the *real* number system.   Before that time real numbers had been built up from the natural numbers by various operations: not only by addition and multiplication but also by, for example, taking logarithms.   Indeed, in 1585 we find Stevin arguing at length that all [real] numbers so constructed are of the same kind.[55]   It was not until about the time of Galois (1811-1832) that there was even a suggestion of considering a system of numbers closed in some way or under certain operations.[56]   So up to that period the real numbers were just taken as some pre-existing numbers. The difficulties in analysis required further investigation of the structure of the real numbers. Dedekind was in the forefront of these investigations, but he also subsequently carried out an analysis of the abstract structure of the natural numbers [1888]. It is that which concerns us here.

In answering his own question, which is the title of his work, *What are numbers and what should they be?*[57] he writes :

Numbers are free creations of the human mind, they serve as a means of apprehending more easily and more sharply the difference of things ([1968], vol.3, p.335).[58]

In a letter to Keferstein (van Heijenoort [1967]) Dedekind presented some account of the development of his ideas on the natural numbers. He also does this to some extent in his preface to [1888]. How he had come to his particular characterization of natural numbers is recorded in the letter to Keferstein. There he makes it clear that he abstracted properties from the sequence of natural numbers. He was not trying to describe them completely, but to give a logically unassailable basis for their theory. He does not discuss the question of their total nature. He writes (in translation):

... the following train of thought, which constitutes the genesis of my essay [Was sind und was sollen die Zahlen?]. How did my essay come to be written? Certainly not in one day; rather, it is a synthesis constructed after protracted labor, based upon a prior analysis of the sequence of natural numbers just as it presents itself, in experience, so to speak, for our consideration. What are the mutually independent fundamental properties of the sequence N, that is, those properties that are not derivable from one another but from which all others follow? And how should we divest these properties of their specifically arithmetic character so that they are subsumed under more general notions and under activities of the understanding without which no thinking is possible at all but with which a foundation is provided for the reliability and completeness of proofs and for the construction of consistent notions and definitions? (op.cit., p.99-100).

Dedekind's analysis consists in regarding the natural numbers as a system (or, as we should say, a set) together with a transformation (or map) of the set into itself satisfying certain properties. At the same time he regards the natural numbers as the simplest kind of infinity. Thus there are two facets to Dedekind's analysis. When Dedekind actually gives his axioms for the natural numbers he gives them in the context of the definition of a simply infinite system. We quote Dedekind, in translation, below. However, we should explain some of Dedekind's terminology first.

As already mentioned "system" means "set". A similar transformation is a one-to-one mapping. If we take such a mapping, ϕ, then the set of elements a, $\phi(a)$, $\phi\phi(a)$, $\phi\phi\phi(a)$, ... is called a chain. So, below, a simply infinite system is a set N with a similar transformation ϕ defined on all elements of N such that $\alpha_. - \delta_.$ hold. Axioms $\gamma_.$ and

$\delta_{\bullet}$ are clear enough. Axiom $\alpha_{\bullet}$ says that ϕ maps N onto a subset N′ of itself, so $\phi(n) \in N$ for all $n \in N$ and N′ = $\{\phi(n) : n \in N\}$. Axiom $\beta_{\bullet}$ says that N is the chain generated from 1, that is N = $\{1, \phi(1), \phi\phi(1),...\}$. Now we can give Dedekind's definition.

*71. Definition. A system N is said to be **simply infinite** when there exists a similar transformation ϕ of N in itself such that N appears as chain (44) of an element not contained in $\phi(N)$. We call this element, which we shall denote in what follows by the symbol 1, the **base-element** of N and say the simply infinite system N is set in order [geordnet] by this transformation ϕ. If we retain the earlier convenient symbols for transforms and chains[59] (IV) then the essence of a simply infinite system N consists in the existence of a transformation ϕ of N and an element 1 which satisfy the following conditions α, β, γ, δ :*

α. N′ $\ni$ N.

β. N = 1_0.

γ. The element 1 is not contained in N′.

δ. The transformation ϕ is similar.

Obviously it follows from α, γ, δ that every simply infinite system N is actually an infinite system (64) because it is similar to a proper part N′ of itself (Dedekind [1888], p.67).

These axioms gave rise to Peano's axioms and Peano duly credits Dedekind (see Peano [1891], p.93). Moreover, Dedekind claims to have come to his analysis from a scrutiny of counting[60] and here emerges a difficulty Dedekind does not seem to recognize. Dedekind has studied counting, but a very important part of his thesis is the ability always to be able to take a successor. Although people have counted arbitrarily far, nevertheless this step marks a departure from counting practice into the realm of pure thought. In a way, Dedekind draws attention to this almost immediately after the above quotation.[61] He writes :

*73. Definition. If in the consideration of a simply infinite system N set in order by a transformation ϕ we entirely neglect the special character of the elements; simply retaining their distinguishability and taking into account only the relations to one another in which they are placed by the order-setting transformation ϕ, then there are these elements called **natural numbers** or **ordinal numbers** or simply **numbers**, and the base-element 1 is called the **base-number** of the **number-series** N. With reference to this freeing the elements from every other content (abstraction) we are justified in calling numbers a free creation of the human mind. The relations or laws which are derived entirely from the conditions α, β, γ, δ in (71) and therefore are always the same in all ordered simply infinite systems, whatever names may happen to be given to the individual elements (compare 134), form the first object of the **science of numbers** or **arithmetic** (Dedekind [1888], p.68).*

Thus he is quite explicitly calling his theoretical constructs *the natural numbers.* Peano writes in a very similar way but he uses a distinctive style. He employs uninflected Latin and a symbolic-abbreviatory presentation. He has a brief half-page explaining his signs and then gives his primitive propositions. We reproduce this below but, as with Dedekind, explain his axioms first. Axiom $\cdot 0$ simply says N_0, the natural numbers, form a class. $\cdot 1$ says 0 is in N_0: Peano· starts his natural numbers from 0 rather than 1. He denotes the successor (the next number) by + so $\cdot 2$ says the successor, a +, of an element of N_0 is again in N_0. $\cdot 3$ is the induction axiom. In modern style it reads: if s is a set such that 0 is in s and, for all a, whenever a is in s then a + is in s, then s contains[62] N_0. $\cdot 4$ says successor is one-to-one on N_0, i.e. a + = b + implies a = b. $\cdot 5$ says a + is not equal to 0. Finally Peano points out that $\cdot 3$ says N_0 is the smallest set which satisfies conditions $\cdot 0$ to $\cdot 2$.

Peano writes:

P_p [= *propositio primitivo*]

$\cdot 0$ $N_0 \in Cls$

$\cdot 1$ $0 \in N_0$

$\cdot 2$ $a \in N_0 \ .\supset. \ a + \in N_0$

$\cdot 3$ $s \in Cls \ . \ 0 \in \underline{s} : \underline{a} \in \underline{s} \supset \underline{a}.\underline{a} + \in \underline{s} : \supset.N_0 \supset \underline{s}$ *Induct*

$\cdot 4$ $a,b \in N_0 \ . \ a + = b + \ .\supset. \ a = b$

$\cdot 5$ $a \in N_0. \ \supset. \ a + - = 0$

.3 N_0 is the smallest class which satisfies conditions
     $\cdot 0 \ \cdot 1 \ \cdot 2; \ldots$

(*Peano [1908], p.27-28*).

He gives no further explanation, merely claiming (*ibid.*, p.28) :

The preceding system P_p suffices for the deduction of all propositions of Arithmetic, Algebra and infinitesimal Calculus.[63]

In our view the approaches just described mark a big change in the study of the natural numbers. Hitherto the natural numbers had been taken for granted, used in counting objects in the world and much theoretical work had been done with a collection of numbers which were, to put it a little crudely, just as sound and real as solid bodies studied in mechanics. Dedekind's and Peano's moves to characterize a structure or system of natural numbers immediately take us into the abstract realm of infinity (as Dedekind himself shows). Here we are dealing with something different, for there is no possibility of verifying whether any infinite collection of objects exists in the universe. At this point we may say we shift from applied mathematics to pure mathematics : from the study of the world by the *use* of mathematics to the study of the world *of* mathematics itself. One cannot argue that Dedekind's natural numbers are identical with the numbers we use in everyday life, for the former have a precise definition, the latter none.[64]

It is, of course, perfectly true that no-one has (yet?) found any discrepancy between results obtained by using the Dedekind-Peano axioms and any obtained by (market-place) calculation.

With Dedekind the system of natural numbers is characterized by a starting point (the axiom that says 1 is a natural number) and a process of going from one number to the (uniquely determined) next number, together with saying that this is the only way to obtain a number. Here there is no clear need for any idea of counting any longer, but only a purely abstract construction.[65] There is, however, a difficulty. For in order to show that at least one system with these properties exists, Dedekind turns to a proof which is remarkably different from the technical proofs which precede it in his book [1912]. He writes:

66. *Theorem.*[66] *There exist infinite systems.*
Proof. My own realm of thoughts, i.e. the totality S of all things, which can be objects of my thought, is infinite. For if s signifies an element of S, then is the thought s', that s can be object of my thought, itself an element of S. If we regard this as transform $\phi(s)$ of the element s then has the transformation ϕ of S, thus determined, the property that the transform S' is part of S ; and S' is certainly a proper part of S, because there are elements in S (e.g. my own ego) which are different from each such thought s' and therefore are not contained in S'. Finally it is clear that if a, b are different elements of S, their transforms a', b' are also different, that therefore the transformation ϕ is a distinct (similar) transformation (26). Hence S is infinite, which was to be proved (Dedekind [1888]. vol.3, p.357).[67]

Thus Dedekind has to stretch the connexion which we have noted earlier between counting words (or symbols) and objects in the world. In the former situation, since people only counted objects that were present, there was no need for number words beyond certain (possibly quite large) limits. Of course in practice (except in games and playful situations, see above, chapter I, §§9,10), we find there are indeed no number words beyond a certain (not necessarily very clearly marked) point. Dedekind seeks an infinite collection in the world, that is, a collection in the world which satisfies his definition of infinite. By "in the world" he understands not only the physical world but also the world of ideas. Here, at least for Dedekind, there is no difficulty about indefinite repetition of the idea of "thought of", whereas it would have been impossible to point to an infinite collection in the physical world or to establish that a collection pointed out was indeed infinite. Further, he did feel that he had produced a sufficient, and sufficiently logical, proof. In his letter to Keferstein he wrote (in translation) :

...does such a system exist at all in the realm of our ideas? Without

a logical proof of existence it would always remain doubtful whether the notion of such a system might not perhaps contain internal contradictions. Hence the need for such proofs (articles 66 and 72 of my essay) (van Heijenoort [1967], p.101).[68]

That Dedekind needs a proof which is as bizarre and un-mathematical as that for his article 66 should perhaps put us on our guard and make us ask what sort of mathematics it constitutes, if any.

In his thesis Brouwer trenchantly criticizes Dedekind for his "proof" in article 66. In translation Brouwer writes :

Let us recall [..] DEDEKIND's famous monograph 'Was sind und was sollen die Zahlen?' ([51], 1888), in which he aims at proving logically the arithmetic of whole numbers, starting from the most primitive notions. For this purpose he constructs a logical system (i.e. a mathematical system of words), the axioms of which are the linguistic images of the connections between the basic notions (whole and part, correspondence between elements, mapping of systems, etc.) and which further is finitely constructed following the logical laws (thus without using complete induction, i.e. the mathematical intuition of 'and so on'). In order to have mathematical significance, this system ought to be completed by a mathematical existence proof. But in order to give that, we shall certainly be forced to use the intuition 'and so on', and then we see at once that we can obtain all the arithmetical theorems much more easily than by DEDEKIND's contrived systems; accordingly DEDEKIND does not give the existence proof. He does give in §66 a proof for: 'Es gibt unendliche Systeme'[69], but 1° a proof is needed for: "Es gibt einfach unendliche Systeme'[70], what is more, and 2° his proof, which introduces 'meine Gedankenwelt'[71] is false, for 'meine Gedankenwelt' cannot be viewed mathematically, so it is not certain that with respect to such a thing the ordinary axioms of whole and part will remain consistent. Consequently DEDEKIND's system has no mathematical significance, in order to give it logical significance, an independent consistency proof would have been needed, but DEDEKIND does not give such a proof ([1975], p.[73]).

Brouwer was writing in 1907, two decades after Dedekind's book was published. Another facet of the natural numbers emerges at this point. We have been primarily concerned with the progression of the natural numbers and even in the context of counting one usually goes one, two, three, etc.. In the mid-nineteenth century Frege presented his analysis of the concept of number. He, too, considered the fundamental rôle of the successor operation (Dedekind's transformation) in, for example, the *Begriffsschrift* (Frege [1879]),[72] of 1879, but he also considered the notion of one-one correspondence.[72] To his work we now turn.

7. Frege

In 1878, Cantor used one-one correspondences for saying when two sets are equivalent, but Frege traces the idea back much further :

Hume[73] *long ago mentioned such a means : "When two numbers are so combined as that the one has always an unit answering to every unit of the other, we pronounce them equal." This opinion, that numerical quality or identity must be defined in terms of one-one correlation, seems in recent years to have gained widespread acceptance among mathematicians*[74] *([1884], p.73).*

At this point in his monograph Frege moves on to the idea of what we now call the equivalence classes generated by such a relation[75] and takes these equivalence classes as his numbers.   Thus he first defines the expression "the concept   F   is equinumerous with the concept   G" (which we shall write as   F ~ G)   by saying there is a one-one correspondence between the objects falling under   F   and those falling under   G.   He next defines "the (cardinal) number [Anzahl] which belongs to the concept   F" (which we shall write as   no(F)) as being "the extension of the concept equal to the concept   F".   Therefore   no(F) = {G : G ~ F}, that is, in modern parlance, the number which belongs to the concept   F   is the equivalence class of   F   under one-one correspondence.

Because of the logical properties of equality Frege is then able to define zero[76].   At this point Frege turns to the idea of "successor" and here we can see the difference from Dedekind's work.

Dedekind had put the question "What are [...] numbers?"[77], while on his first page Frege asked "what the number one was"[78].   So although Dedekind had defined numbers in terms of one and successor[79], Frege wants to push the analysis further.   He writes :

Now we have already decided in favour of the view that the individual numbers are best derived, in the way proposed by LEIBNIZ, MILL, J. GRASSMANN and others, from the number one together with increase by one, but that these definitions remain incomplete so long as the number one and increase by one are themselves undefined ([1884], p.25).

Frege's further analysis of "successor" takes him into the idea of series, building on his earlier ideas in the *Begriffsschrift* (see Frege [1879], p.57 f.).   Since Frege is concerned now only with the particular sequence of natural numbers, he needs only to define successor here.   So what he says is :

The proposition: "there exists a concept   F, and an object falling under it   x, such that the Number which belongs to the concept   F   is   n and the Number which belongs to the concept 'falling under   F   but not identical with   x' is   m " is to mean the same as "n   follows in the series of natural numbers directly after   m." (Frege [1884], p.89).

In a terser way we might write

$$\exists F \; \exists x (F(x) \; \& \; no(F) = n \; \& \; no(F \text{ but} \neq x) = m).$$

It thus appears at first sight as though Frege has set up the machinery for counting up to any particular number to which there corresponds that many objects in the world.   He explicitly draws attention to the need for a justification of a number greater than any previously obtained.

There is nothing in this so far to state that for every Number there exists another Number which follows directly after it, or after which it directly follows, in the series of natural numbers (ibid., p.91).

And then :

Now in order to prove that after every Number (n) in the series of natural numbers a Number directly follows, we must produce a concept to which this latter Number belongs (ibid., p.92).

But since Frege believes that truths of mathematics (or logic) are eternal, or, perhaps better, atemporal truths, whether a number follows another or not "has in general absolutely nothing to do with our attention and the circumstances in which we transfer it ; on the contrary, it is a question of fact" (Frege [1884], p.93).   So Frege does not even have to appeal to our world of thoughts as Dedekind does.   (On the other hand, he does not say here just what a fact is or how we know a fact.)

Having thus, as he puts it, made it possible "to reduce the argument from   n   to   (n+1), which on the face of it is peculiar to mathematics, to the general laws of logic" ([1884], p.93), he has of course taken the natural numbers from the realm of worldly or applied mathematics into the purely abstract world - just as Dedekind did.   Again, therefore, we have a detachment of the idea of natural number from counting.   Frege has done one further thing.   He has put in sharp relief the separate questions of the existence of each (natural) number.   Frege also treats the induction axiom, though in a somewhat different form (see [1879], proposition 69.)   And after the items discussed above he moves on smoothly, very smoothly, to deal with infinite numbers in his very next section.

From Frege and Dedekind we therefore have a picture of each individual number and the whole range of natural numbers.   Dedekind shows how starting from   1   and a one-one transformation ϕ one gets a unique system, that is, if   *   and ψ constitute another system satisfying the Dedekind-Peano axioms, then the two systems are isomorphic under the correspondence :

$$1 \quad \leftrightarrow \quad *$$

$$\phi(1) \quad \leftrightarrow \quad \psi(*)$$

$$\phi\phi(1) \quad \leftrightarrow \quad \psi\psi(*)$$

$$\vdots$$

Frege shows that there is a unique equivalence class containing both $\{1,\phi(1),..., \quad \phi...\phi(1)\}$ and $\{*,\psi(*),..., \quad \psi...\psi(*)\}$ where there are the same number of ϕ's and ψ's in the parentheses: a concept which precedes the notion of number, as we have seen, and Frege was at pains to point out.

Russell puts this very nicely in his [1911] where in talking about cardinal numbers (that is, numbers under Frege's definition) he comments :

The cardinal numbers which are increased by the addition of unity are the same as the inductive numbers, that is, the numbers which we call **natural**. *This proposition identifies the two definitions of the infinite. We could say that a class is infinite whenever it contains a part that we could put in one-one relation with the entire class; according to this definition, a number is infinite whenever it is not increased by the addition of unity. Or else we can say that we call* **finite** *whole number any number which obeys complete induction starting from zero. That is to say : let us call* **recurrent property** *any property which holds for* n + 1 *given that it holds for* n, *and let us call* **inductive property** *any recurrent property which holds for zero. Then we shall call* **inductive** *number any number which possesses all inductive properties, that is to say, any number for which proofs by means of complete induction are valid. It is easy to see that the inductive numbers are the same as the natural numbers[80] 0, 1, 2, ..., 100,..., 1000, ..., and that (admitting the axiom of infinity) there are numbers which are not inductive numbers, for example, the number of inductive numbers. The so-called* **principle** *of complete induction becomes therefore a definition, namely, the definition of inductive numbers. Then we can say that the infinite numbers are the non-inductive numbers. (See Russell [1911], p.169.)*

However, to say that it "is easy to see that the inductive numbers are the same as the natural numbers 0, 1, 2, ..., 100,..., 1000,..." leaves the nature of the natural numbers quite unclear.

What appears to be happening is that the unsophisticated idea of natural numbers leads, on further investigation, to a number of different idealized notions. These notions are formulated with the precision of mathematical language and therefore, by their very nature, cannot agree precisely with the unsophisticated and inexact one. It is certainly interesting, and important for mathematics, that various sharply formulated notions coincide but the divergence from our intuitive ideas needs to be recognized.[81]

The Dedekind-Peano axioms led, with the growth of mathematical logic in the first half of this century, to an almost exclusive concern with the formal axioms. Indeed, attention even shifted from the Dedekind-Peano axioms to an even more formal set of axioms in the formal language of the predicate calculus. It was Skolem who in [1934] showed that these formal axioms admitted interpretations different from the natural numbers, the so-called "non-standard models" of arithmetic. In 1931 Gödel showed that any reasonable axioms must necessarily be deficient in another sense. Namely, if the axioms are consistent (that is, do not lead to a contradiction), then there are sentences which are true but cannot be proved from the axioms. To put it another way : if the axioms are consistent then there are statements S such that neither S nor the negation of S is provable from the formal axioms. For a long time attempts have been made to produce "natural" examples of such statements. (Gödel's original ones were highly contrived and based on the idea of the liar paradox "This sentence is false" though they were not, in fact, paradoxical at all.) In his [1977] Paris exhibited a rather straightforward mathematical statement having Gödel's property. Thus at last the way is open to continue the development of our ideas of the natural numbers beyond what appeared to be Frege's impasse of perfection. We have seen how the natural numbers have developed. We expect that development to continue.

8. Conclusion

We have tried to follow the development of the natural numbers from their emergence first on artifacts and then in language. We have noted at least three stages. One is the abstraction of number and of specific numbers ("two" for example) from the surroundings, a second is the development of a basic system of counting (for example on fingers or with heaps of shells) and a third is the idea of induction and the arbitrarily long continuation of the natural number sequence.

In each of these stages (which merge into each other) there is a human contribution. For the first stage objects have to be distinguished from their environment and mentally arranged. For the second, correspondences between objects or between objects and actions are required. For the third, the idea that if we can count *so* far we can *always* count further leads to the notion of mathematical induction. Each of these stages may well have taken centuries : certainly the last one has connexions with Euclid and its development to an explicit and formal notion has been variously dated to start almost anywhere from the fourteenth century on.

Throughout there have been increases of precision in concepts and the use of words but these have depended on human activities and needs. The myth of a final explanation of numbers held sway for a time but now the processes of development have been set in action again.[82] With Dedekind we may at least partly agree, though we may wish to go no further than to say that "numbers are free creations of the human mind ; they serve as a means of apprehending more sharply the difference of things" ([1968], vol.3, p.335; English tr. in Beman [1963].

Part 2

Complex Numbers

The modern mind is conditioned with certain concepts which did not exist in the ancient mind, and the historian's task is not only to get rid of these concepts but to base his reasonings and substantiations on the knowledge of ancient works (Lam & Shen [1986], p.18).

In Part 1 we witnessed the very slow development of the natural numbers from primitive counting through to a formal set of axioms for the natural numbers. In this part we see a quite different phenomenon. Relatively speaking, complex numbers erupted onto the mathematical scene.

As students, most of us first learnt of complex numbers in the solution of quadratic equations $ax^2 + bx + c = 0$, where $b^2 < 4ac$. To us, consequently, they seem to be inherent in such equations. In history, however, quadratic equations, or at least problems equivalent to them, were known for between 1500 and 4500 years before complex numbers and when complex numbers were at last introduced it was in the context of the solution not of quadratic equations, but of cubic equations. It is true, however, that from a theoretical point of view they arise from the solution of a quadratic equation which forms a step in the solution of a cubic.

This discovery, or invention, of complex numbers, — I do not wish to discuss the philosophical issue of whether it was a discovery or an invention — seems to be due to Bombelli in his book (1572) which he wrote around 1550. It is true, however, that Cardano's book published in 1545 (see [1968]) contains a particular pair of complex numbers. In introducing complex numbers Bombelli writes:

I have found a new kind of tied cube root [of the form $\sqrt[3]{a+\sqrt{b}}$] very different from the others ([1572], p.133).

In chapter III we shall trace the path to the introduction of complex numbers and in particular examine why they were not introduced earlier. This will require us to look at the development of equations and of what was meant by a solution to an equation. From the point of view we are adopting in this book, these form an integral part of the developments which led up to the introduction of complex numbers.

In chapter IV we shall consider the work of Bombelli, his contemporaries and successors. We shall see that the introduction of complex numbers was quite pragmatic. Bombelli showed that his cases of complex numbers worked, that is, were effective, in that they did give correct solutions to his problems. We shall conclude this part of the book with Leibniz's demonstration that the formula which Bombelli used does, in fact, always give correct results: a nice result whose genesis also led to results which were surprising even after complex numbers had been known and used for a hundred years.

56

We shall not here go into the nineteenth century justification of complex numbers and their reduction to pairs of real numbers given by Hamilton ([1967], p.97-100), nor shall we treat the geometric presentation of complex numbers which developed after Leibniz. Treatments of these can be found in many books on the history of mathematics (e.g. Smith [1959], vol. I, p.55-66; Novy [1973], p.117-121).

Chapter III

Latency

Nowadays complex numbers are introduced in the context of solving quadratic equations. Some authors claim that quadratic equations have been solved for at least two millenia. Thus Heath ([1925], vol.I, p.383 and vol.II, p.263) and van der Waerden ([1975], p.121) assert either that certain propositions of Euclid *are* solutions of quadratic equations or lead to such solutions. It has also been suggested that the Babylonians solved quadratic equations (van der Waerden [1975], p.69). Euclid flourished about 300 B.C., the Babylonian tablets referred to are at least a hundred years earlier (see van der Waerden [1975], p.69 f.). Yet it is also well known that Cardano was the first (in 1545 [1968]) to publish any work containing any complex numbers. Why, when it is so obvious to us that complex numbers come into the solution of quadratics, was it not clear to anyone before the sixteenth century?

1. The Babylonians

In order to see what happened we have to follow the principles stated by Lam and Shen in the quotation at the beginning of this part of our book. We have to rid our minds of modern mathematical concepts and try to understand how people thought in the past about quadratic equations and what were the concepts they used.

Many books on the history of mathematics do not do this but are content to describe the ancient mathematics in terms of modern concepts. Kline [1972] is an excellent example. He makes it easy for the modern reader to understand *what* was done. However, when we want to see the mechanism of development, that approach can be quite unhelpful and even misleading. Thus van der Waerden ([1975], vol.I, p.69) asks:

How may the Babylonians have obtained the solution of the quadratic equations $x^2 \pm ax = b$*?*

The simple answer is that they would not have had the slightest idea what the question meant! The notion of "equation", let alone that of "quadratic equation", did not exist at that time. The resolution of questions which, in modern parlance, give rise to quadratic equations developed on two fronts. On the one hand the approach of the Greeks was geometrical. On the other there was a more algorithmic[1] or recipe approach followed first by the Babylonians and other people at the eastern end of the Mediterranean and even more practically in China. We shall see that the geometric and algorithmic-algebraic approaches came together gradually, culminating in work of the Arabs towards 1000 A.D.. This concerted treatment is an explicit goal of Omar Khayyam (d. 1131) who is, today, much better known for his love poem, the *Rubai'yat*.

58

In the next few centuries these Arab gains were transmitted, with some loss, back into Europe, through trade and the Arab penetration of the western Mediterranean.   One major source for the Arabs, namely the work of Diophantos, was not transmitted directly by the Arabs to the West but was rediscovered in Greek manuscript by Regiomontanus (about 1460)[2] and Bombelli (about 1545)[3] and became a major inspiration for algebraic work.

At that time, the first half of the sixteenth century, there was a mathematical climate in which many mathematicians were anxious to find an algebraic solution of, that is, a recipe for solving, cubic equations.   A solution was found and in the process complex numbers were introduced.

The recipe solution of problems *does* go back to the Babylonians. First recall that the Babylonians used the scale of sixty rather than the decimal scale, thus 53 means $5 \times 60^1 + 3 \times 60^0$.   For clarity we put commas between the digits, thus 5, 3 and use a semi-colon for the sexagesimal equivalent of the decimal point.   Thus 14,30;15 means

$$14 \times 60^1 + 30 \times 60^0 + 15 \times 60^{-1} = 870\tfrac{1}{4}$$

in our notation.

Corresponding to our solution $x = \sqrt{\{(b/2)^2 + c\}} + b/2$ of $x^2 - bx = c$, we find the following treatment of what we would now write as the solution of

$$x^2 - x = 14, 30 \quad \text{(that is, } x^2 - x = 14\tfrac{1}{2}\text{).}$$

The solution given proceeds as follows:

Take 1, the coefficient (of x).   Divide 1 into two parts.
0;30 × 0;30 = 0;15, you add to 14,30 and 14,30;15 has the root 29;30.   You
add to 29;30 the 0;30 which you have multiplied by itself, and 30 is (the
side of) the square (Neugebauer [1935], vol.III, p.6).

This is Neugebauer's translation.   The solution is by the traditional method of the completion of the square.   However, since all Babylonian problems have answers and all such answers are, necessarily, positive - though perhaps it is less misleading to say "unsigned" - because they are usually lengths, the situation where a problem could produce a complex number as an acceptable answer simply does not occur.

Here we see the Babylonians giving a solution of one problem.   The same *technique* is applied time and again (see Neugebauer [1935], vol.III, *passim*). The Babylonians are using a recipe for solving problems equivalent (for us) to quadratic equations.   There is no justification of the method by, for example, a geometric proof.   The method itself is one which proceeds by means of the basic operations of addition, subtraction (with a non-negative answer), multiplication and division plus the extraction of (square) roots.   We shall call this type of solution a *radical* solution.

Since the Babylonians did extract (approximations to) some irrational square roots, the possibility of an irrational root is specifically included here.

Similarly in China in the *Jiuzhang suanshu* ("Nine Chapters on the Mathematical Art", of about 100 B.C. - 100 A.D.), we find algorithmic procedures for solving quadratic equations, but all the problems considered have solutions in their context, that is, positive real number solutions (see Li and Du [1987], p.53). Again, these solutions usually represent lengths.

2. Euclid

When we turn to the books of Euclid we find an entirely different approach. Here it is geometry and proof which dominate. Later Western mathematicians, instead of using arithmetic and algorithms, based their solutions of quadratic equations on four propositions of Euclid, but these propositions do not present, or even represent, solutions of quadratics. The four propositions are Book II, Propositions 5 and 5, and Book VI, Propositions 28 and 29. The propositions are as follows (in Heath's translation:

Book II, Proposition 5.

If a straight line be cut into equal and unequal segments, the rectangle contained by the unequal segments of the whole together with the square on the straight line between the points of section is equal to the square on the half (Heath [1925], I, 382).

Book II, Proposition 6.

If a straight line be bisected and a straight line be added to it in a straight line, the rectangle contained by the whole with the added straight line and the added straight line together with the square on the half is equal to the square on the straight line made up of the half and the added straight line (Heath [1925], I, 385).

Book VI, Proposition 28.

To a given straight line to apply a parallelogram, equal to a given rectilineal figure and deficient by a parallelogrammic figure similar to a given one: thus the given rectilineal figure must not be greater than the parallelogram described on the half of the straight line and similar to the defect (Heath [1925], II, 260).

Book VI, Proposition 29.

To a given straight line to apply a parallelogram equal to a given rectilineal figure and exceeding by a parallelogrammic figure similar to a given one (Heath [1925], II, 265).

Now, Book II of Euclid is, apart from one or two propositions about triangles at the end, concerned with problems of the following type. Given a line cut it at a particular point and then investigate the area generated by taking the two segments of the line as sides of a rectangle.

Thus we can paraphrase II.5 in modern algebraic notation as

$$ab + (a - \tfrac{1}{2}(a+b))^2 = (\tfrac{1}{2}(a+b))^2.$$

However, a more instructive procedure is to follow Euclid more closely. Draw a figure and then consider the equation or its equivalent read from left to right.

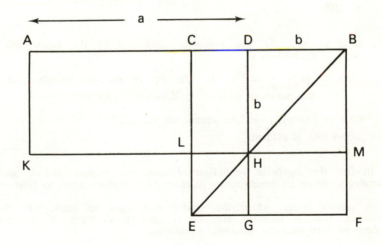

In this way the proposition is rendered much more clearly *visible* (Heath [1925], vol.I, p.382).

Let us write ⊓CH to mean the area of the rectangle whose diagonal is CH and ∆LHE to mean the area of the triangle with vertices L, H and E. First we show that ⊓CH = ⊓HF.

⊓CH + ∆LHE + ∆DBH = ∆CBE

= $\tfrac{1}{2}$⊓CF

$\quad$ = $\triangle$BFE

$\quad$ = $\square$HF + $\triangle$HGE + $\triangle$BMH = $\triangle$BFE

while $\quad\quad$ $\triangle$LHE = $\frac{1}{2}\square$LG = $\triangle$HGE $\:$ and similarly $\triangle$DBH = $\frac{1}{2}\square$DM = $\triangle$BMH.

Subtracting these two items from the first and fourth lines establishes $\square$CH = $\square$HF.

Next $\quad\quad$ $\square$AH + $\square$LG

$\quad$ = $\square$AL + $\square$CH + $\square$LG

$\quad$ = $\square$AL + $\square$HF + $\square$LG $\:$ (since $\:\square$CH = $\square$HF)

$\quad$ = $\square$CM + $\square$HF + $\square$LG $\:$ (since $\:$C$\:$ is the midpoint of $\:$AB)

$\quad$ = $\square$BE.

Now $\quad\square$AH $\:(=\:ab)$ $\quad$ is "the rectangle contained by the equal and unequal
$\quad\quad\quad\quad\quad\quad$ segments",

$\quad\quad\square$LG $\:(= [a - \frac{1}{2}(a+b)]^2)$ $\quad$ is "the square on the straight line between
$\quad\quad\quad\quad\quad\quad$ the points of section" andf section" and

$\quad\quad\square$BE $\:(= (\frac{1}{2}(a+b))^2)$ $\:$ is "the square on the. half"
so the proposition is evident.

In fact the algebraic symbolism obscures the picture and it also tempts the reader to bring an anachronistic mathematical sophistication to bear.

A similar figure, which the reader may care to draw, will show how Proposition II.6 is likewise clarified - and how there is no reason to introduce or even mention quadratic equations.

When we turn to Euclid VI.28 and 29 we find Euclid is concerned with constructing particular areas. This whole book of Euclid is concerned with constructing lines and figures satisfying certain constraints regarding proportionality or size.

The problem in VI.27 is to construct a point $\:$S$\:$ and a line $\:$TR parallel to $\:$AB$\:$ such that the parallelogram $\:$TS$\:$ is equal (in area) to a given area $\:$C$\:$ while the parallelogram $\:$SR$\:$ is similar to the given parallelogram $\:$D. $\:$ Now the area of the parallelogram $\:$TB$\:$ is proportional to the side $\:$AT$\:$ since $\:$AB$\:$ is fixed. $\:$ But the area of the parallelogram $\:$SR$\:$ is proportional to the square of $\:$AT$\:(=\:RB)\:$ and therefore proportional also to the square of $\:$SB. $\:$ So we need to find a length $\:$AT$\:$ such that parallelogram $\:$AT$\:$ by $\:$AB$\:$ minus the parallelogram (with sides parallel to AT and AB but of size)AT by AT equals area $\:$C.

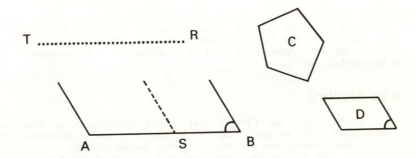

In this form the move to a quadratic equation is clear for us without, we hope, doing too much injury to Euclid's presentation.   Of course, one must have the difference of the areas of the two parallelograms positive to make sense for Euclid and must therefore satisfy the condition stated in Proposition VI.28.   (In modern terms this is the condition that the discriminant $b^2 - 4ac$ be positive.   However, it would, and does, take a certain amount of effort to work out what the coefficients actually are (cf. Heath [1925], vol.II, p.263) - a task that Euclid never could have considered.)

When we turn to Euclid's *Data*, Proposition 85 ([1883-1916], p.167-169), we find the solution of the problem: Given lengths AB and AΓ and the area of the rectangle with sides AB and AΓ, find AB and AΓ.

The present-day interpretation of this is presented as solving a quadratic equation, the sum and product of whose roots is given.

This latter presentation and the one from Euclid's *Elements* both preclude the consideration of situations where quadratic equations could have negative or zero roots.   Note also that the construction and the proof which Euclid gives preclude the consideration of complex roots.   Such considerations simply cannot be formulated in the given environment. Moreover, since his concern is geometry, the required solution of a problem is a geometric one: in particular a line.

This is not the same as saying that the Greeks admitted irrational solutions.   For, as many authors have pointed out and as we shall see, in chapter V of this book, the correlation of a number with every length on a line of given numerical length was a notion which caused great consternation to the Greeks.   Thus, for example, Dedron & Itard say:

63

*The Greeks never identified a non-rational square root with a number
([1974], vol.2, p.102).*

Thus the Babylonians gave rules to compute numbers using square roots
(radical solutions) and the Greeks gave geometric solutions to problems.
The Chinese also gave rules for computing numbers. When we come, much
later in time, to the Arabs, we find a combination of numerical and
geometric approaches.

3. Hindu Algebra

Further to the east, the Chinese and Hindu mathematicians developed
algebraic techniques from an early date. It was the Hindus, however, who
developed what has become our algebra. Hindu mathematicians are believed
to have been the first to introduce zero and they also used negative numbers
from an early date. By the time of Brahmegupta (b. 598, d. after 665,
(D.S.B.)), the rules for algebraic operations appear almost as axioms.
Brahmegupta writes (in Colebrooke's translation):

*Rule for addition of affirmative and negative quantities and cipher:
... The sum of two affirmative quantities is affirmative; of two negative is
negative; of an affirmative and a negative is their difference ...
(Colebrooke [1817], ch. XVIII, §II, p.339).*

Brahmegupta follows this by rules for subtraction, multiplication,
division and also for manipulating surds. In particular, as an example, he
divides by $\sqrt{18} + \sqrt{3}$. He does this essentially by multiplying dividend and
divisor by $\sqrt{18} - \sqrt{3}$ to rationalize the denominator since
$(\sqrt{18} + \sqrt{3})(\sqrt{18} - \sqrt{3}) = 18 - 3$ (Colebrooke [1817], p.342). He follows this
with the solution of quadratic equations.

*Rule: To the absolute number multiplied by the [coefficient of the]
square, add the square of half the [coefficient of the] unknown, the square
root of the sum, less half the [coefficient of the] unknown, being divided
by the [coefficient of the] square, is the unknown (Colebrooke [1817], Ch.
XVIII, IV, p.347).*

There follows the example

$$x^2 - 10x = -9$$

-9 times $1 + (\frac{1}{2}(-10))^2$ gives 16. $\sqrt{16} = 4$. $4 - \frac{1}{2}(-10) = 9$. 9 divided
by $1 = 9$ the unknown.

This is the Babylonian method modified by the admission of negative
quantities. Indeed, both Brahmegupta and Bhaskara (born a long time after
him in 1115 A.D.) used negative quantities freely. Bhaskara writes:

*The square of an affirmative [i.e. positive] or of a negative quantity
is affirmative; and the root of an affirmative quantity is two-fold,*

*positive and negative.   There is no square root of a negative quantity; for
it is not a square (Colebrooke [1817], p.135.)*[4]

So he may be viewed as dismissing, or at least missing, imaginary
numbers.   He then continues:

*Example. .. Say promptly likewise what is the root of nine affirmative
and negative, respectively? ... Statement: 9. Answer: The root is 3 or 3̣
[= -3].   Statement: 9̣ [= -9].   Answer: there is no root, since it [-9] is
not a square.*

It is curious that this aspect of Hindu algebra did not pass to the
Arabs.   Neither did the Hindu use of abbreviations for the unknown.
Brahmegupta uses the initial letters of words for colours as his unknowns
employing several in a single problem (Colebrooke [1817], p.348).   I know
of no Arab up to and including al-Karkhi (see below) who used such
abbreviations.[5]

Nevertheless, it is generally accepted that algebra, including the
radical solution of quadratic equations, was transmitted from the Hindus to
the Arabs before the ninth century had passed.

4. Al-Khwārizmi

Already under Harūn al-Rashīd, who reigned from A.D. 786-809, al-Hajjāj
had brought an Arabic translation of Euclid's *Elements* back from Europe (see
Gandz [1932], p.65), so Euclidean knowledge was available to the Arabs. This
knowledge was not always used but much mathematics developed under the
Caliphs.   Caliph al-Ma'mun, who reigned in Baghdad from 813-833,
established a "House of Wisdom" (Bayt-al-Hikma) and it was to him that
al-Khwārizmi dedicated his *Algebra* of 830.   Abū Ja'far Muhammad ibn Mūsā
al-Khwārizmi was born before 800 and died after 847 (D.S.B.).   He appears,
from his book, not to have known of Diophantos who preceded him in time (see
below) but may have got his algebraic information either from the Hindus
directly,   presumably because of trade links with the East, or via the Jews.
Gandz shows convincingly that the geometry in al-Khwārizmi's book comes from
the "practical" book, the *Mishna ha Middot*, probably written about 150 A.D.
by Rabbi Nehemiah.   In this case Gandz ([1932], p.64) suggests that he
learnt of the Jewish work via either a Persian or a Syrian translation of
this *Mishna*.

Al-Khwārizmi's treatise of 830 is called *Hisāb al-jabr w'al-muqābala*.
It was translated into Latin twice in the twelfth century by Gerard of
Cremona and Robert of Chester (c. 1140) under the title *Liber algebrae et
almucabala*.   *Algebra* is used today by surgeons to mean *bone-setting*, i.e.
the restoration of bones, and the idea of restoration is present in the
mathematical context, too.   It refers here to the idea of completing, in
particular, a square, as the quotations below show.   *Almucabala* (we use the
Latin form of the word) refers to the reduction (by cancellation) of the
powers of the unknown (with positive coefficients) on both sides of the

equation.[6] For example, the move from $50 + x^2 = 29 + 10x$ to $21 + x^2 = 10x$, or similarly from $50x + x^2 = 29x + 10x^2$ to $21x + x^2 = 10x^2$. al-Khwārizmi did give the rules for manipulating powers up to the second but he did not use symbols and all his work is written out in words. Like everyone else until the sixteenth century, al-Khwārizmi always uses specific numerical examples, but he has moved towards a general classification. Already in the *Vi'ja-Gan'ita* of Bhāskara, quadratic equations were separated from linear equations. However, because the Hindus used negative quantities, all quadratics (with real roots) could be treated in the same way.[7] This was not the case for al-Khwārizmi. He had to have positive or "zero" coefficients and by "zero" coefficients we mean, of course, missing terms. From his [1831][8] we obtain the following list of six types of equation of the second degree:

Squares equal to roots, squares equal to numbers, roots equal to numbers,... a square and roots equal to numbers, a square and numbers equal to roots, and roots and numbers equal to a square.[9]

For the justification of his methods al-Khwārizmi uses results known to Euclid. This may explain why the seventh type: $x^2 + ax + b = 0$, where a, b are positive, does not arise. Indeed, there is no Euclidean geometric picture for this case. Nevertheless, at the beginning of his *Algebra* al-Khwārizmi says:

When I considered what people generally want in calculating, I found that it always is a number.

I also observed that every number is composed of units, and that any number may be divided into units.

Moreover, I found that every number, which may be expressed from one to ten, surpasses the preceding by one unit: afterwards the ten is doubled or tripled, just as before the units were: thus arise twenty, thirty, &c., until a hundred; then the hundred is doubled and tripled in the same manner as the units and the tens, up to a thousand; then the thousand can be thus repeated at any complex number; and so forth to the utmost limit of numeration.

I observed that the numbers which are required in calculating by Completion and Reduction [al-jabr w'al muqabala] are of three kinds, namely, roots, squares, and simple numbers relative to neither root nor square.

A root is any quantity which is to be multiplied by itself, consisting of units, or numbers ascending, or fractions descending.[10]

A square is the whole amount of the root multiplied by itself.

A simple number is any number which may be pronounced without reference to root or square.

A number belonging to one of these three classes may be equal to a number of another class; you may say, for instance, "squares are equal to roots," or "squares are equal to numbers," or "roots are equal to numbers" ([1831], p.5-6).[11]

Thus al-Khwārizmi expresses himself in terms of numbers without reference to surfaces.

Next, however, we find a technique which appears in Bhāskara's *Vi'ja-Gan'ita* (e.g. §146 f. in Colebrooke [1817], p.222 f.). Namely, the presentation of an algebraic (radical) rule together with a geometric demonstration. In Bhāskara we find (Colebrooke [1817], §147):

Rule: Twice the product of the upright and side[12], being added to the square of their difference, is equal to the sum of their squares, just as with two unknown quantities.[13] Hence for facility, it is rightly said 'The square root of the sum of the squares of upright and side, is the hypotenuse?'[14] Placing the same portions of the figure in another form, see[15]

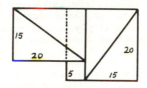

(Colebrooke [1817], p.222-223).

This is of course our old friend Pythagoras' theorem.

Al-Khwārizmi proceeds similarly, giving first the radical rule and then the geometric demonstration. We quote his results for $x^2 + 21 = 10x$.

Squares and Numbers are equal to Roots;[16] for instance, 'a square and twenty-one in numbers are equal to ten roots of the same square'. That is[17] to say, what must be the amount of a square, which, when twenty-one dirhems are added to it, becomes equal to the equivalent of ten roots of that square? Solution: Halve the number of the roots; the moiety is five. Multiply this by itself; the product is twenty-five. Subtract from this the twenty-one which are connected with the square; the remainder is four. Extract its root; it is two. Subtract this from the moiety of the roots, which is five; the remainder is three. This is the root of the square which you required, and the square is nine. Or you may add the root to the moiety of the roots; the sum is seven; this is the root of the square which you sought for, and the square itself is forty-nine. When you meet with an instance which refers you to this case, try its solution by addition. And if that do not serve, then subtraction certainly will. For in this case both addition and subtraction may be employed, which will not answer in any other of the three cases in which the number of the roots must be halved. And know, that, when in a question belonging to this case you have halved the number of the roots and multiplied the moiety by itself, if the product be less than the number of dirhems connected with the square, then the instance is impossible[18]; but if the product be equal to the dirhems by themselves, then the root of the square is equal to the moiety of the roots alone, without either addition or subtraction ([1831], p.11-12).

Here again we see that he considers such problems impossible when, as we say, $b^2 < 4c$ for an equation $x^2 + c = bx$. This prohibition ties in with his geometrical justifications. Thus in considering a square and 10 roots equal 39 numbers ($x^2 + 10x = 39$, a favourite problem then and for centuries after) he says:

Demonstration of the Case: "a Square and ten Roots are equal to thirty-nine Dirhems".[19]

The figure to explain this is a quadrate, the sides of which are unknown. It represents the square, the which, or the root of which, you wish to know. This is the figure AB, each side of which may be considered as one of its roots; and if you multiply one of these sides by any number, then the amount of that number may be looked upon as the number of the roots which are added to the square. Each side of the quadrate represents the root of the square; and, as in the instance, the roots were connected with the square, we may take one-fourth of ten, that is to say, two and a half, and combine it with each of the four sides of the figure. Thus with the original quadrate AB, four new parallelograms are combined, each having a side of the quadrate as its length, and the number of two and a half as its breadth; they are the parallelograms C, G; T, and K. We have now a quadrate of equal, though unknown sides; but in each of the four corners of which a square piece of two and a half multiplied by two and a half is wanting. In order to compensate for this want and to complete the quadrate, we must add (to that which we have already) four times the square of two and a half, that is, twenty-five. We know (by the statement) that the first figure, namely, the quadrate representing the square, together with the four parallelograms around it, which represent the ten roots, is equal to thirty-nine of numbers. If to this we add twenty-five, which is the equivalent of the four quadrates at the corners of the figure AB, by which the great figure DH is completed, then we know that this together makes sixty-four. One side of this great quadrate is its root, that is, eight. If we subtract twice a fourth of ten, that is five, from eight, as from the two extremities of the side of the great quadrate DH, then the remainder of such a side will be three, and that is the root of the square, or the side of the original figure AB. It must be observed, that we have halved the number of the roots, and added the product of the moiety multiplied by itself to the number thirty-nine, in order to complete the great figure in its four corners; because the fourth of any number multiplied by itself, and then by four, is equal to the product of the moiety of that number multiplied by itself.[20] Accordingly, we multiplied only the moiety of the roots by itself, instead of multiplying its fourth by itself, and then by four. This is the figure:

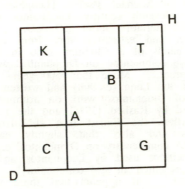

(al-Khwārizmi [1831], p.13-15).

There are vague echoes of Euclid here but no clear indication that al-Khwārizmi knew Euclid directly, if at all. Indeed, Gandz believes the echoes are inaudible:

Euclid's elements in their spirit and letter are entirely unknown to [al-Khwārizmi]. Al-Khowārizmī has neither definitions, nor axioms, nor postulates, nor any demonstration of the Euclidean kind. He has just a plain treatise on mensuration [as the geometric part of his Algebra] ... And even this is not his own but an almost verbal [sic] translation of an old Hebrew geometry ([1932], p.64).[21]

Thus al-Khwārizmi, although he may not have been very original in his *Algebra*, wrote a book which was widely used, in Arabic as well as in Latin, as a source of information on linear and quadratic equations and effectively maintained a veto on the solution of problems which give rise to quadratic equations with negative discriminant.

Relatively soon after this, another influence began to take effect. The expansion of Arabic science, which Caliph al-Ma'mun had initiated, continued. By the year 1000 there were at least two Arabs who had worked on the arithmetic of Diophantos. Although Diophantos lived before al-Khwārizmi, the latter makes no mention of Diophantos. Now we turn to Diophantos' work.

5. Diophantos

According to a letter of Michael Psellus (Diophantos [1893-95], vol.II, p.35 f.), who lived in the eleventh century, Diophantos of Alexandria lived at the time of Anatolius and therefore about 250 A.D. We do not have a more precise date for him and estimates vary from about 150-350 A.D. Hypatia, who died at the hands of a Christian mob in 415, is reported by Suidas (c. 1000 A.D.) to have commented on Diophantos' work but there is no copy of this commentary extant. More than 500 years later al-Nadim (*fl.* 987-988) reported that Qustā ibn Lūqā (c. 900) had written a "Commentary on three and one half books of Diophantos' work on arithmetical problems" and this work has been re-edited by Rashed [1975] and the new edition reviewed by Sinaceur [1976]. Further, the Fihrist (see al-Nadim [1871-2]) of 987 A.D. mentions Diophantos and also that, slightly earlier, Abu'l Wafā (940-988) had written both "A commentary on Diophantos' algebra" and a "Book on the proof of the propositions used by Diophantos and of those that he himself [Abu'l Wafā'] has presented in his commentary". Neither of these last two has been preserved. It appears that these works derived their information from at least one manuscript in Constantinople. Perhaps there was a manuscript related to the one recently edited by Sesiano [1982]. Despite this attention from the Arabs, Diophantos' works remained unknown to the Europeans till nearly 1500 A.D..

We have eleven books of Diophantos' works: ten of what are believed to have been thirteen[22] books of the *Arithmetic* and *On polygonal numbers*. The first book of the *Arithmetic* commences with notation for squares and cubes going up to cubocubes or sixth powers.

In the Greek manuscripts Diophantos uses an abbreviation which looks like a final sigma s and Δ^Y, K^Y for $\delta \upsilon \nu \alpha \mu \iota s$, $\kappa \upsilon \beta o s$ (power [= square], cube) going on to $\Delta^Y \Delta$, $\Delta^Y K$ and $K^Y K$ for the fourth, fifth and sixth powers, that is, square-square, square-cube, cubo-cube ([1893-95], vol. I, p.5). There is no geometric reticence here. He immediately turns to the inverses of powers and follows with rules for multiplication of such powers and inverses, giving a list of all products of these involving only powers, inverses and answers with exponent ≤ 6.

One gets the impression from some versions of Diophantos that he is using a notation surprisingly close to modern usage. However, there are variations between the manuscripts and I believe that the abbreviations (in particular □ for square and ∧ for minus[23]) are merely scribes' normal abbreviations or late introductions. Certainly the Arabs who did have translations of Diophantos did not employ abbreviations at all (see Sesiano [1982]). But then since many Arabic words only comprise three consonants in their usual written form, the need for abbreviations would be felt less. Thus Diophantos does not have a calculus but at most a small set of abbreviations.

He gives the rules for minus:

Minus multiplied by minus gives plus, minus times plus gives minus and the sign of minus is an incomplete ψ *pointing down,* ∧ *(Diophantos [1893-95], p.12).*[24]

Eventually we shall see how Bombelli and others isolated the rules "minus times plus equals minus", "minus times minus equals plus" (see below, Chapter IV, §1).   Note that here Diophantos treats multiplication first, as did Bombelli later, though there is, as is usual with Diophantos, no justification provided.

Then Diophantos gives rules for rearrangement which are echoed by the Arabs and all later workers in algebra.

Now for starting this discipline it is well to begin by adding, subtracting and multiplying expressions as an exercise; and we know how to add positive and negative expressions not of the same coefficient to other expressions which are positive or similarly positive and negative.

Finally if it happens in a problem that some expression is equal to another not having the same coefficient it is necessary to subtract the same from the same until one expression [term] becomes equal to one expression [term].   If an expression [term] occurs in one of the two expressions [sides of the equation] as negative then it is necessary to add the negative expression [term] to both sides until the expressions [terms] on both sides become positive and again to subtract the same from the same from both sides until one expression [term] [on one side only] remains ([1893-95], p.14).

Having outlined his (algebraic) procedures, Diophantos then goes on to say (in translation):

Apply this method to the equations of the problem as far as possible until a single expression is left equal to another single expression. Later I will show you how to solve the problem where two expressions [added together] remain equal to a single expression (ibid.).

Diophantos then goes on to the problems.   He starts with linear problems working up to other polynomial equations.   One noteworthy feature is the reduction of each problem to one involving a single unknown (though he does sometimes employ a second, auxiliary, unknown).

To find two numbers in a given ratio, such that their difference is also given.   Suppose the larger be the quintuple of the smaller and their difference is 20 units.   Let the smaller number be 1 unknown; then the larger will be 5 unknowns ([1893-95], p.21, Bk. I, 6 [= Chapter IV]).

Nowhere does Diophantos give geometric justifications of his results, nor is he constrained by consideration of dimensions.[25]   Nevertheless, as we have already pointed out, he does not consider negative solutions of problems at all.   Thus in Book I, Chapter XIV, we find:

To find two numbers such that their product has a given ratio to their sum.[26]   It is necessary to suppose that the number of units in one of the numbers be greater than the number corresponding to the given ratio.[27,28]

71

Thus in his general procedures (see above), he is prepared to manipulate quantities affected by minus even though he does not allow negative quantities as solutions.

Diophantos has thus, to a significant extent, overcome the Euclidean requirement that only those dimensions which correspond to possible ones in the Euclidean world be used. It is interesting to note that Diophantos, unlike Viète at the end of the sixteenth century (see below, §IV.4), would freely add e.g. lengths to areas and was not constrained to preserve homogeneity of dimension. Nevertheless, Diophantos was surely aware of Euclid's work and was under his influence when he considered the problem:

To find two numbers such that their sum and the sum of their squares are equal to given numbers. It is necessary that twice the sum of the squares of the numbers exceed the square of the sum of the numbers by a square ([1893-95], p.63, Bk. II, κη).[29]

In his treatment, therefore, of these problems he appears much closer to our algebraic approach than to the Euclidean model.

This then was material which was almost certainly available to the Arabs by, at the latest, 1000 A.D.. It is to be noted that by this time they already had rules for finding solutions using root extraction (what we have called "radical solutions") as well as geometric demonstrations in the style of al-Khwārizmi, but they developed much more than that. Diophantos' problems were used and considerable progress was made leading to the definitive work of Omar Khayyam, which we shall shortly consider.

Omar Khayyam is one of two particularly famous Arab mathematicians who lived beyond 1000 A.D.. The other was Abu Beqr Mohammed Ben Alhaçan al-Karkhī (al-Karkhī for short), who is also known as al-Karajī. He died between 1019 and 1029 (D.S.B.) and wrote an algebra book dedicated to the Vizier and called by a derivative of the Vizier's name: *al-Fakhri* (al-Karkhī [1853]).

al-Fakhri contains a version of much of Diophantos' books II, III and IV and starts off just as does Diophantos by listing powers of جذر, jidr, the thing or root = $\alpha\rho\iota\theta\mu o\varsigma$ of Diophantos (see Sesiano [1982], *t*.24, p.284, S. 802, p.318). He lists them exactly as Diophantos does, going up to quadrato-cubo-cube (the eighth power), مالكعب كعب and then saying (in Woepcke's. translation) "and so on to infinity"[30] (al-Karkhī [1853], p.48). Throughout this, and all the rest of the book, al-Karkhī, like all the other Arabs, avoids abbreviations or symbols. He gives rules for the arithmetic operations including (essentially) the multiplication of polynomials. Like all the Arabs he eschews negative quantities throughout but remarks that for multiplication "it is necessary here to count negative quantities (المقادير المستثنا ة) as terms"[31] (al Karkhī [1853], p.51).

al-Karkhī usually gives a numerical example for his rules but he does not give any sort of proof beyond geometric pictures. Often he explicitly says he is giving a solution in the style of Diophantos (على مذهب ديو فنطس) ([1853], p.66). He does not treat equations above the second degree, except for ones which can easily be reduced to at most second degree equations followed by the extraction of roots.[32]   The solutions of quadratics are based explicitly on the Euclidean theorems we noted above (e.g. Euclid II, 5, see [1853], p.63).

6. Omar Khayyam

By the time we come to the work of Omar Khayyam, or Ghiyath Uddin (or al-Din) Abu'l Fath Umar ibn Ibraham al-Khayyami (or al-Nisaburi), to give him his full name, considerable progress had been made on the solution of certain cubic equations.   Omar Khayyam, who was born about 1048, probably in Nishapur, and died in Nishapur, Persia, on 4 December 1131, wrote "Commentaries to difficulties in the Introduction to Euclid's Book" (see D.S.B.).   He was familiar with other Greek mathematical books, in particular Apollonius' *Conic Sections*.   Although al-Khwarizmi had only dealt with quadratic equations, Omar Khayyam considered and contributed to the solution of cubic equations.   He did this in a style which is admirably described by Isaac Newton:

The Ancients, as we learn from Pappus, initially attacked the trisection of an angle and the finding of two mean proportionals by way of the straight line and circle, but to no effect.   Subsequently, they began to take numerous other lines — such as the conchoid, cissoid and the conic sections — into consideration and by means of certain of these they solved those problems.   At length, having pondered the matter more deeply and accepted conic sections into geometry, they distinguished problems into three types: plane ones, solvable by lines — namely, the straight line and circle — deriving their origin from the plane; solid ones, solved by lines deriving their course from consideration of a solid — a cone, to be exact —; and linear ones, for whose solution more complicated lines were required. Following this distinction it is alien to geometry to solve solid problems by any lines other than conics, especially if no other lines except the straight line, circle and conics are to be accepted into geometry (Newton [1972], p.423).

In pursuing the solution of cubic equations Omar Khayyam blends solid geometry and the geometry of conics.   He makes the claim:

But whenever cubes   [x^3]   come in, and among them and other places there is an equation, we need solid geometry, and especially conics and conic sections because a cube [x^3] is a solid (Amir-Moez [1961], p.329).

Omar Khayyam was the first to give a method for solving *all* such equations (Eves [1958], p.285).   However, he does base his work heavily on proposition II.5 of Archimedes' *The Sphere and the Cylinder* (Heath [1897]) whose solution is given by a method equivalent (for both Omar Khayyam and us) to solving a cubic equation.   This proposition is not included in Omar Khayyam's manuscript but does feature in an addition to that manuscript in a note by Ibn al-Haitham (see al-Karkhī [1853], p.92 f.).   Woepcke notes:

It appears that this lemma in a quite unusual way seized the attention of the Arab geometers.   Since Archimedes had not given a solution to it, it is perhaps the case that there was a point of honour in proving that they knew how to surmount, easily, an obstacle which seemed to have stopped Archimedes (Omar Khayyam [1851], p. xiij).[33]

In fact, Eutocius (b. 480 A.D., see D.S.B.) believed he had found Archimedes' solution, albeit in a mutilated state, and had reconstructed the solution which can now be found in Heath's version of Archimedes [1897], Proposition II.4. The problem is to determine where to cut a sphere by a plane so that the volumes of the two parts of the sphere thereby produced shall be in a given ratio.

Before treating _cubics, however, Omar Khayyam classifies quadratics in the style of al-Khwārizmi. He then gives a numerical solution and a geometrical solution for all the possible cases.[34] (Remember that non-positive and complex solutions are impossible for the Arabs and that $ax^2 + bx + c = 0$ is not considered with a, b, c all non-negative.) But this is of considerably less interest for us than his full treatment of cubics.

In this treatment he started from the solution of the Archimedean problem mentioned above. He wanted to give both a geometric and a numerical solution for all the cubic equations he could consider. By a "numerical solution" he meant a positive integral solution, but such he does not give. He says: "The demonstration for these six kinds is only possible by using the properties of conic sections" (Omar Khayyam [1851], p.11).[35]

He points out that only isolated equations had been treated before (*ibid.*). Indeed, in another paper (Amir-Moez [1961]) he refers to Aboo [*sic*] Nassre ibn Aragh, who came from Khawarazm (presumably the same place as Mohammed al-Khwarizmi). Aboo Nassre ibn Aragh

... tried the problem of Archimedes about the construction of the side of a heptagon in a circle ...   He also used notations of algebraists. Consequently the analysis lead [sic] to **cubes** *and* **squares** *equal to numbers. He solved this equation with conic sections.   There is no doubt that this man was a great mathematician (Amir-Moez [1961], p.331).*

Omar Khayyam also cites Aboo-al-Wojood:

The problem that Aboo Sohl Koohi, Aboo-al-Wafa Bozejani, Aboo Hamed Saghani, and their friends in Bagdad at Azd-ed-Doleh court were not able to solve is: We want to divide ten into two parts so that the sum of the squares of them plus the ratio of the larger to the smaller will be equal to seventy.   The analysis of this problem leads to **squares** *equal to* **cubes,** **objects,** *and numbers.*[36] *These learned men were perplexed about this problem for a long time until Aboo-al-Wojood solved it and it was treasured in the library of the Kings of Samani (Amir-Moez [1961], p.331).*

Omar Khayyam was not aware of any attempt to solve cubics by ancient foreign mathematicians.

Ancient mathematicians of other languages have not discovered these ideas and nothing has reached us through translation to our language (Amir-Moez [1961], p.330).

But some cubics were solved:

Thus these equations are three forms of polynomials but two of them are of three terms and one of four terms.   There is only one singleton[37] which is the equation of cube and numbers.   These equations were solved by learned men before us, but we have not received any of the other forms with details of their work (Amir-Moez [1961], p.331).

And towards the end of that paper he promises:

If the opportunity arises and I can succeed, I shall bring all of these fourteen forms with all their branches and cases, and how to distinguish whatever is possible or impossible so that a paper, containing elements which are greatly useful in this art will be prepared (Amir-Moez [1961], p.331).

After these remarks Omar Khayyam goes on to establish a solution of the equation $x^3 + 200x = 2000$, using a conic.

In his *Algebra* Omar Khayyam fulfils his promise.[38]   He proceeds to reduce the problem to the construction of two conics.   Of course his solution is always a line and not an expression involving radicals.   Thus at the end of the earlier work he writes:

In this case the line DB [the solution] is the same as AL of the previous problem and we have proved it is known in value. By saying it is known in value I do not mean that its magnitude is known because these two ideas are different.   By being known in value I mean what Euclid meant in the book of Constructions.   That is, we can construct a magnitude equal to it (Amir-Moez [1961], p.33-334).

And we observe that in his work Omar Khayyam has constantly been using magnitudes rather than numbers, for he sets this problem up in terms of lines and gives geometric prescriptions.   That he quite definitely considers arbitrary magnitudes is clear from an earlier remark:

... Archimedes has said that two lines AB and BC are known in values and are connected to each other in one direction.   The ratio of BC to BE is known.   Then, according to what has been said in the book of Constructions, CE is known (Figure).

Then he has said that we choose the ratio of HC to CE as the ratio of the square of AB to the square of AH.   But he has not said how this idea is done.   This problem requires conic sections (Amir-Moez [1961], p.330).

Thus when we said above that he considered problems that we classify as ones concerning equations with positive coefficients, he was not restricted to rational coefficients but allowed any given positive magnitudes. This is quite clear in the *Algebra*, where for example in the case of $x^3 + a = bx$ (in present-day notation) he writes in translation:

> *Let the line AB be equal to the side of a square equal [in area] to the number of roots [b], and construct a solid having as base the square on AB, and equal [in volume] to the given number (Omar Khayyam [1851], p.34).*[39]

We do not find anything like these considerations again until Bombelli.[40] But what Omar Khayyam was doing was using old Greek knowledge in a new way in order to solve cubics. By the nature of the theory of conics and geometry at that time, this way could not lead to complex roots of cubic equations, for there was no way in which they could be drawn in his figures. Curiously, however, it could have led to a consideration of negative roots if, in appropriate cases, both branches of a hyperbola had been drawn: a point on which Woepcke takes Omar Khayyam to task ([1851], p.xvj). But Omar Khayyam did try to investigate the situation of when the conics would fail to intersect. Woepcke includes later work which improved Omar Khayyam's results (see Omar Khayyam [1851], p.96 f.). Omar Khayyam did already show, for example, that for the equation $x^3 + a = cx^2$ (*op.cit.*, p.40 f.), where a and c are positive, that if $\sqrt[3]{a} \geq c$ then by considering $x = \sqrt[3]{a}$, $x < \sqrt[3]{a}$ and $x > \sqrt[3]{a}$ there are no positive roots. However, al-Qūhī, the unnamed author (see Omar Khayyam in D.S.B.) of the addition B to Omar Khayyam's *Algebra* (*op.cit.*, p.97) shows that for $x^3 + a = cx^2$ the limiting case corresponds to $4c^3 = 27a$ (cf. *op.cit.*, p.xviij). Thus the discriminant of the cubic begins to loom on the horizon.[41]

By the end of Omar Khayyam's life, Arab science was being transmitted to Europe. Presumably this was connected with trade. For example we have:

> *Beginning in 1133, the commune of Pisa concluded a series of treaties with the Almoravid and Almohad sultans of Morocco and Tunis, which had the ultimate effect of ensuring a regular and mutually profitable commerce between the two peoples (Thomson [1975], p.2, footnote omitted).*

7. The Early Italians

Through contact with the Arab world some knowledge of algebra was transmitted to Europe and indeed, as a direct consequence of the trade connexions, Leonardo of Pisa (Fibonacci) (born *ca.* 1170, died after 1240) lived in Bougie, Algeria, for some time. He also travelled around the Mediterranean, for in (Leonardo Pisano [1857, 1862], vol. I, p.1, we read:

> *When my father, who had been appointed by his country as public notary in the customs at Bugie acting for the Pisan merchants going [i.e. trading] there, was in charge, he summoned me to him while I was still a child, and*

having an eye to usefulness and future convenience, desired me to stay there and receive instruction in the school of accounting. There, when I had been introduced to the art of the Indians' nine symbols [i.e. the Hindu-Arabic numerals] through remarkable teaching, knowledge of the art very soon pleased me above all else and I came to understand it, for whatever was studied by this art in Egypt, Syria, Greece, Sicily and Provence, in all its various forms. [42]

In 1202 he wrote the first version of his *Liber Abbaci* (Leonardo Pisano [1857, 1862], vol.I). This is principally concerned with the use of Arabic numerals – that is, with algorism as it became known in the West. In this book he also treats (as we would say) simultaneous linear equations (*op.cit.*, vol.I, p.227).

Square roots do not come in until near the end (*op.cit.*, vol.I, p.353). However, there he treats surds (explicitly $\sqrt{10}$) giving both a numerical approximation (by fractions) and a geometric construction

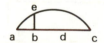

$$be^2 = ab.bc.$$

(*op.cit.*, vol.I, p.353). Later he treats quadratic equations in the style of the Arabs, going on to such problems (*op.cit.*, vol.I, p.415) as become, in our modern notation, $x_1 + x_2 = 10$, $x_1^2 = 32x_2$. Woepcke (al Karkhī [1853], p.25 f.) gives a close analysis of the similarities between many of the problems and solutions in Fibonacci and al-Karkhī. It appears that Fibonacci got his techniques and some problems from al-Karkhī or, at the very least, a similar Arab source.

Leonardo's talent for working with fractions and extracting (approximate) roots leads in his later work *Flos* (Leonardo Pisano [1857, 1862], vol.II, p.227-247) to a solution of an equation $10x + 2x^2 + x^3 = 20$ (p.228 f.). He goes on to show, geometrically, that the solution must be neither (positive) integral, nor rational, nor a square root of a rational. He gives an approximate solution (working in Roman numerals and using sexagesimal notation):

And because it was not possible to solve this question in any other of the above ways [i.e. is not rational etc.] I worked to reduce the solution to an approximation. And I found one from the ten nominated roots, that is the number ab, according to the approximation is one plus 22 minutes, 7 seconds, 42 thirds, 33 fourths, 4 fifths and 40 sixths. [43]

That is $1°22'7''42'''33^{iv}4^{v}40^{vi}$ in sexagesimal notation, or 1.368808 in decimal notation. All this is done by geometric construction using rectilinear figures (not solids) and appealing to Euclid Book VI (p. 229). In other writings he goes on to Diophantine-type problems and here he uses Arabic numerals but still geometric determinations (*Liber quadratorum compositus a leonardo pisano Anni M.CC.XXV.* [Leonardo Pisano [1857, 1862], vol.II, p.253 f.), though he has referred five years earlier to the mixing of arithmetic and geometry in his mathematics (*op.cit.*, vol.I, p.1).

After Fibonacci there were a number of Italian mathematicians working in algebra but we do not know of much progress. Indeed, Cossali ([1797], I, p.17) says that there was only a slow diffusion of algebra after Fibonacci. However, there were a lot of people working in algebra (see van Egmond [1977]). By the end of the fifteenth century Fra Luca Pacioli had published a book (Birkhoff & MacLane [1941]) containing a lot of well-known material. For example, on f. 117r he shows how to simplify $\sqrt{24} + \sqrt{6}$ to $\sqrt{54}$ by using the figure[44]

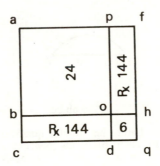

He also devotes a great deal of space to more difficult calculations with square roots, etc.. What is interesting here is that he lays out his calculations in the margin in a very modern way, using $\bar{p}$ for *più* (= plus), $\bar{m}$ for *meno* (minus) and the square root sign R . For example, in the margin of f. 137v we find

$$\nu^{a} \; . \; 6 \; . \; \bar{p} \; . \; R \; . \; 10$$

$$18 \; . \; \bar{m} \; . \; R \; . \; 90$$

$$\overline{108 \; . \; \bar{m} \; . \; R \; .3240. \bar{p} \; . \; R \; . \; 3240 \; . \; \bar{m} \; . \; R \; . \; 90}$$

$$hoc \; est \; .78$$

which (*pace* the error of a missed final   0   because of no space being available to the printer) transcribes as:

$$multiply\ 6 + \sqrt{10}$$
$$18 - \sqrt{90}$$

$$108 - \sqrt{3240} + \sqrt{3240} - \sqrt{900}$$
$$This\ is\ 78$$

The multiplication of the square roots is achieved, as in our previous example, by relying on Euclid.   (This use of symbols is not present in Fibonacci's work.)   Pacioli goes on to calculate with repeated roots writing, for example, R_xR_x160 for $\sqrt{\sqrt{160}}$.

Algebra is introduced on f.144r with a flourish:

The great art: that is, the speculative practice: otherwise called Algebra and almucabala in the Arabic language:[45]

He gives rules for solving quadratics, giving geometric demonstrations using Euclid (e.g. f. 146r):

The verse of the first rule: If the unknown and the square together are equal to a number [$bx + x^2 = c$], having taken half [the co-efficient of] the unknown you must produce its square and add this to the number [$(b/2)^2 + c$]: take all of which by a root [$\sqrt{((b/2)^2 + c)}$] from half the [co-efficient of the] unknown and the side of the square is recovered [$x = \sqrt{((b/2)^2+c)} - b/2$].[46]

But he does not treat cubics at all, though he does mention generalizations to (what we write as) $x^4 = a + bx^2$ saying one can go on to infinity by pursuing this method.   He notes that $x^4 + ax^2 = bx$ and $x^4 + ax = bx^2$ are "imposible" (f. 149r), a note which is only lightly touched on at the end of this part (f. 150r), where he says he is unable to give general rules.

Other Italians also used radical expressions in the solution of polynomial equations in the thirteenth century (see van Egmond, [1983], p.399-421).

From the Italians the French had learnt much and Nicolas Chuquet (living 1484) and his pupil Estienne de La Roche (otherwise known as Villefranche) produced some interesting algebra.

Before 1880 it was La Roche who had had the finer reputation, having published the first edition of his *Arithmetic* (La Roche [1520], in 1520, with a second edition in 1538.   Nicolas Chuquet's work remained unknown until a manuscript was discovered and then published in 1880 by Marre [1880].   There it was announced that most of La Roche's book was a direct

79

copy of the Chuquet manuscript. Thereby Chuquet's reputation was established and La Roche's over-turned. Recently Flegg, Hay and Moss [1985] have published a translation of substantial parts of Chuquet's manuscript, thereby making it readily accessible.

In fact, La Roche (on the second unnumbered folio *recto* of [1520]) gives credit to

nicolas chuquet parisien: philippe friscobaldi florentin: frere luques de burgo.[47]

La Roche says that he is writing

with only a little addition to what I have found and experimented with in my time in practice: and of all this I have made a little tract entitled The Arithmetic of Estienne de La Roche...[48]

La Roche's credit-giving is at least as meagre as others' about that time (Cardano, Bombelli, for example) but he does acknowledge his debt. In fact Chuquet and La Roche were almost certainly deriving their material from Italy where

A host of works on arithmetic and algebra, written on manuscripts, belong to this same epoch ... (Marre [1880], p.560).[49]

Chuquet not only used $\bar{p}$ and $\bar{m}$ for plus and minus and R for root but introduced superscript notation writing R^2 for square root, R^3 for cube root and so on. Various forms of bracketting or ligature also came in. Thus on p. 728 of (Marre [1880]) we find

$$R^2 . R^2 . 7 . \bar{m} . 2 \text{ [multiplied by]}$$
$$\overline{R^2 . R^2 . 7 . \bar{p} . 2} \text{ [gives] } R^2 . 3$$

that is, $\sqrt{(\sqrt{7}-2)}$ times $\sqrt{(\sqrt{7}+2)} = \sqrt{3}$. However, he also used the superscript notation for powers and got as far as writing a polynomial fraction

$$\frac{30 . \bar{m} . 1^1}{1^2 . p . 1^1} \text{ that is } \frac{30 - x}{x^2 + x}$$

(Marre [1880], p.745).

Marre's adulation of Chuquet isnot diminished by the evidence of Chuquet's sources:

From the beginning of his "Triparty" Nicolas Chuquet shows the Italian provenance of the science which he expounds with a precision and a clarity which is totally French (Marre [1880], p.566).[50]

La Roche does in fact go quite a bit further than Chuquet. He has used the notation $R^{\square}$ for cube root as well as R^3 (f. 29v), R and R^4

for fourth roots, going on to R^5, R^6, R^7, etc. He spends a long time on the multiplication of tied roots ("racines liées"), that is, roots of expressions already involving roots (see f. 39r, for example). He uses a symbol for the unknown and deals with expressions involving an unknown at length (see f. 45) treating, for example, expressions which may be transcribed as

$$\frac{20x + 20\square}{10\square + 2x + 1} \quad i.e. \quad \frac{20x + 20x^2}{10x^2 + 2x + 1}$$

where I have kept La Roche's $\square$ but modernized the other symbols. He also writes 12^1, 12^2, 12^3, 12^4 (on f. 42) for our $12x$, $12x^2$, $12x^3$, $12x^4$. (This notation is introduced by Chuquet: see Marre [1880] p.742.) La Roche's multiplications are set out like Pacioli's.

8. Bologna

This new surge of interest was more strongly felt in Italy. In Bologna about that time the mathematicians (maestri d'abbaco) were involved in solving numerical problems and had to hold public debates. On the results of these

> depended not only their reputation in the city or in the University, but also tenure of appointment and increases in salary (al-Nadīm [1871-2]. p.25). Disputations took place in public squares, in churches and in the courts kept by noblemen and princes, who esteemed it an honour to count among their retinue scholars skilled not only in the casting of astrological predictions, but also in disputation on difficult and rare mathematical problems (ibid.).[51]

Pacioli was the only mathematician of note in the late fifteenth century in Bologna (Bortolotti [1974], p.24), but his work had an enormous diffusion and was the basis for all the sixteenth century work in mathematics in Italy (ibid., p.28). We shall consider several mathematicians who built on this foundation.

We shall treat Scipione dal Ferro, Cardano, Bombelli and, very briefly, Tartaglia (Tartalea) and Ferrari. (Later we shall also look at the work of non-Italians such as Stevin, Viète and Descartes.)

> Scipione dal Ferro (1465-1526) was one of five joint holders of the arithmetic and geometry chair [at Bologna] from 1496 - 1526 (Rose [1976], p.145).

He probably met Pacioli when the latter was lecturing at Bologna in 1501-2 (ibid.). It is generally stated that Scipione was the first person to solve general cubic equations (see e.g. Cardano [1968], p.96 and D.S.B., dal Ferro). However, it would appear there is some confusion here. Scipione dal Ferro gave a rule[52], which we discuss below, and this *rule* is applicable to all cubic equations (even $ax^3 + bx^2 + cx + d = 0$). But at that time cubic equations were classified in the same way as Omar Khayyam classified them (though there is no evidence that the Italians knew of Omar Khayyam's work). That is to say, by phrases such as

cube and things [unknowns] equal to a number

In our notation this is $ax^3 + bx = c$ (e.g. Cardano [1968], ch. XI f.). Thus, in particular, $ax^3 + bx^2 + cx + d = 0$ was not considered when a, b, c, d were all positive.

It is not known how Scipione discovered his rule, but Bombelli in the manuscript of his *Algebra* [1572], published in 1572, says that Scipione was adept in manipulating expressions involving cube roots and square roots (see also Bortolotti [1974], p.44) and indeed showed how to manipulate fractions involving the sum of three cube roots in the denominator in order to obtain a denominator free of cube roots. In Tartalea's *Quesiti XXV* of Book Nine of (Masotti [1959], p.108), Tartalea solves certain cubics of the form $x^3 + px^2 = q$ by trying to find a solution of the form $\sqrt{a} - b$ and equating rational and irrational parts (p, q being positive integers). The D.S.B. entry for dal Ferro suggests that this idea might have been used by Scipione in obtaining his rule. Scipione had indeed written out his work, for Ferrari wrote in the second of his *Cartelli* on the calends of April (1 April) 1547:

Four years ago when Cardano was going to Florence and I accompanied him, we saw at Bologna Hannibal Della Nave, a clever and humane man who showed us a little book in the hand of Scipione dal Ferro, his father-in-law, written a long time ago, in which that discovery was elegantly and learnedly presented (Bortolotti [1923], p.387).[53]

Although the rule is generally applicable *now* (in its algebraic form), it seems very likely that Scipione only discovered it for equations of the form $x^3 + px = q$. In the contest in which Scipione's rule was employed, Scipione's pupil Antonio Maria Fior could not solve all (the then considered) types of cubic equation but his opponent Tartalea could.

There is, as it happens, a note in a manuscript in the University of Bologna (MS. 595N) which reports "Dal Ferro's rule for the solution of cubic equations".[54] It is headed *From the Cavaliere Bolognetti, who had it from the Bolognese master of former days, Scipion dal Ferro.*

On unknowns and cubes equal to numbers.[55]

It goes on to give a solution, according to the usual rule, of $3x^3 + 18x = 60$.[56]

This page lends support to the view that Scipione's as opposed to Tartalea's (see below), rule was intended only for the case $x^3 + ax = b$.

Tartalea reports in his [1554] *Quesito XXV* (f. 107r) (see Masotti [1959]) a conversation he had with Maestro Zuanne.

*MASTER ZUANNE.   I have heard that some time ago you entered into a
disputation with Master Antoniomaria Fior and that in the end you reached
agreement whereby he was to propound 30 problems for you, each of a
different kind, set down in writing and sealed, to be deposited with Master
Per Iacomo di Zambelli, notary and similarly you would propound 30 problems
for him, each of a different kind also.   This you both did, fixing a term
of 40 or 50 days for each of you to solve these problems and agreeing that
whichever of you within that time should be adjudged to have solved the
greater number of the questions you had been given would take the honours,
together with some small reward suggested for each problem.   And it has
been reported to me, and confirmed by Fina a Bressa, that you solved all 30
of his [problems] in the space of two hours.   I find that hard to believe!*

*NICOLO.   All that you have been told or had reported to you is true.   And
the reason why I was able to solve his 30 [problems] in so short a time is
that all 30 concerned work involving the algebra of unknowns and cubes
equalling numbers.   [He did this] believing that I would be unable to solve
any of them, because Fr. Luca asserts in his treatise that it is impossible
to solve such problems by any general rule.   However, by good fortune, only
eight days before the time fixed for collecting from the notary the two sets
of 30 sealed problems, I had discovered the general rule for such
expressions (Masotti [1959], f.107).[57]*

This was on 12 February 1535 (1534 Venice dating).   Tartalea shortly
after the above wrote:

*... and from a few hints and casual data arising out of that discovery
I found next day a further general rule for problems [of the form] unknowns
and numbers equalling cubes [i.e. $ax + b = x^3$] (Masotti [1959], f.107v).[58]*

*All the problems I propounded for him were [indeed] each of a different
kind.   I did this in order to show him my versatility, and that my
grounding lay not merely in one or two, or even three, private discoveries
of mine, or in secrets, although I had kept them to myself for greater*

*safety.   Moreover, I could have set him another 1000, not just 30, instead,
as agreed, I propounded all [30] each of a different kind, to show that I
thought little of him and had no cause whatever to fear him (ibid.).[59]*

Tartalea found it hard to remember his own rules for the three cases,
so he composed a rhyme as he reports in *Quesito XXXIII* (see Masotti[1959],
f.124r).   The following verse translation (by R.G. Keightley) gives the
spirit (bearing in mind that we should understand "[coefficient of the]
unknown" in lines 5 and 15).

*In cases where the cube and the unknown
Together equal some whole number, known:
Find first two numbers diff'ring by that same;
Their product, then, as is the common fame,
Will equal one third, cubed, of your unknown;
The residue of their cube roots, when shown*

And properly subtracted, next will give
Your main unknown in value, as I live!
As to the second matter of this kind,
When cube on one side lonely you shall find,
The other terms together being bound:
Two numbers from that one, once they are found,
Together multiplied, swift as a bird,
Give product clear and simple, of one-third
Cubed of th' unknown; by common precept, these
You take, cube-rooted; add them, if you please,
T'achieve your object in their sum with ease.
The third case, now, in these our little sums,
From the second is solved; for, as it comes,
In kind it is the same, or so say I!
These things I found — O, say not tardily! —
In thrice five-hundred, four and thirty more,
Of this our age; the gallant proof's in store
Where City's girt by Adriatic shore.[60]

Scipione's rule may be presented, in modern guise, as follows, though we consider only equations with no square term. (Eliminating the square term appears to have been well-known. For example, Cardano gives such rules in Chapter XVIII f. of (Cardano [1968]) and so does Bombelli [1572], p.333 f.).

So consider $x^3 + ax = b$. Now $(u - v)^3 = u^3 - 3uv(u - v) - v^3$. So

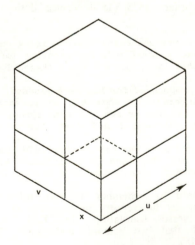

if we put $u = x + v$, then $u = x + v$, $x^3 = u^3 - 3uvx - v^3$ and therefore $x^3 + 3uvx = u^3 - v^3$. Comparing this with $x^3 + ax = b$ it suffices to find numbers u, v such that $3uv = a$, $u^3 - v^3 = b$. The numbers u, v[61] are then the cube roots of U and -V where U and V are the solutions[61] of U + V = b, UV = -(a/3). U, V can easily be found by Euclidean methods (using proposition 85 of the *Data* (Euclid [1962]).

84

Cardano promised he would keep the method secret but it has often been said that Cardano in fact published it in his *Ars Magna* in 1545.   It would appear that Cardano felt justified in publishing both the rule (which constitutes the method) and the proof, because the proof was his own work even though the rule was not.   He did give Tartaglia some credit but only to the same sort of extent as La Roche gave to Chuquet.   Cardano writes in *Ars Magna* (in Witmer's translation, Cardano [1968], p.96):

Scipio Ferro of Bologna well-nigh thirty years ago discovered this rule and handed it on to Antonio Maria Fior of Venice, whose contest with Niccolò Tartaglia of Brescia gave Niccolò occasion to discover it.   He [Tartaglia] gave it to me in response to my entreaties, though withholding the demonstration.   Armed with this assistance, I sought out its demonstration in [various] forms.   This was very difficult.   My version of it follows.

Cardano then proceeds to give his geometric proof.   He proves

$$(u + v)^3 = u^3 + v^3 + 3uv(u + v).$$

Cardano, like almost all his predecessors, thought of cubes as solids, so his proof is Euclidean in style.   However, he does consider problems of the fourth degree.   He gives the impression of someone who is capable of doing something, namely operating with unfamiliar objects (biquadratics, complex numbers, for example), while not being clear just what is going on. This impression is confirmed by the first record we have of a point where complex numbers could be (naturally) introduced.   Bortolotti quotes a letter of Cardano to Tartaglia:

The first report concerning the difficulties created by the appearance of these new numerical entities was published in a letter from Cardano to Tartaglia, dated 4 August 1539 and reproduced in XXXVIII of **Quesiti   et Inventioni diverse di N. Tartaglia** *(1546): 'I have sent to enquire (Cardano writes) after the solution to various problems for which you have given me no answer, one of which concerns the cube equal to unknowns plus a number $(x^3 = px + q)$.   I have certainly grasped this rule, but when the cube of one-third of the [coefficient of] the unknown is greater in value than the square of one-half of the number (when $p^3/27 > q^2/4$), then, it appears, I cannot make it fit into the equation.   However, I would esteem it a favour if you would solve this problem for me: "1 cube equal to 9 unknowns plus 10" $(x^3 = 9x + 10)$.'   Here Cardano was asking, in essence, what meaning should be attributed to an expression of the form*

$$\sqrt[3]{\frac{q}{2} + \sqrt{\frac{q^2}{4} - \frac{p^3}{27}}} + \sqrt[3]{\frac{q}{2} - \sqrt{\frac{q^2}{4} - \frac{p^3}{27}}},$$

$$\text{when } \frac{q^2}{4} < \frac{p^3}{27}.$$

*Tartaglia is unwilling to admit to the embarrassment caused to him by such a
problem and pretends that he purposely tried to lead Cardano in the wrong
direction.   'I propose to see (he tells us) whether I can perhaps alter the
data he possesses, that is, turn him away from the right track and make him
take some other...'   He therefore writes to him: 'And thus I say in reply
that you have not mastered the true way of solving problems of this kind,
and indeed I would say that your methods are totally false' (Bortolotti
[1923], p.286-387).*[62]

Ars Magna took five years of Cardano's life, so it is not surprising
that in six years his views have not changed much (as evidenced by this
letter and the book itself) (Cardano [1968], p.261).

Just how ambiguous the idea of complex numbers was comes over well in a
passage which leads into the often-quoted passage from Chapter XXXVII of *Ars
Magna*:

*So progresses arithmetic subtlety the end of which, as is said, is as
refined as it is useless (Cardano [1968], p.220).*

We prefer, however, to consider the whole passage where Cardano is
giving the *demonstration* that two numbers whose sum is   10   and whose
product is   40   are   5p : Rm : 15   and   5m : Rm : 15   or, in our notation, 5
+ √-15   and   5 - √-15.   Cardano writes:

DEMONSTRATION

*That the true significance of this rule may be made clear, let the line   AB
[see Figure], which is called   10, be the line which is to be divided into
two parts whose rectangle is to be 40.   Now since 40 is the quadruple of
10, we wish four times the whole of AB.   Therefore, make   AD   the square
on   AC, the half of   AB.   From   AD   subtract four times   AB.   If there is
a remainder, its root should be added to and subtracted from   AC   thus
showing the parts [into which   AB   was to be divided].   Even when such a
residue is negative, you will nevertheless imagine   √-15   to be the
difference between   AD   and the quadruple of   AB   which you should add to
and subtract from   AC   to find what was sought.   That is   5 + √25-40   and
5 - √25-40, or   5 + √-15   and   5 - √-15.   Dismissing mental tortures, and
multiplying 5 + √-15 by   5 - √-15, we obtain   25 - (-15)   which is   +15.
Therefore the product is   40.   However, the nature of   AD   is not the same
as that of   40   or AB   because a surface is far from a number or a line.
This, however, is closest to this quantity, which is truly puzzling since
operations may not be performed with it as with a pure negative number or
with the other numbers.   Nor can we find it by adding the square of half
the number to the product number and take away and add from the root of the
sum of half of the dividend.   For example, in the case of dividing   10
into two parts whose product is   40, you add   25, the square of one half of
10, to   40   making 65.   From the root of this subtract   5   and then add   5
and according to similar reasoning you will have   √65 + 5   and   √65 - 5.
But these numbers differ by   10, and do not make   10   jointly, but   √260,
and thus far does Arithmetical subtlety go, of which this, the extreme, is,
as I have said, so subtle that it is useless.*[63]

For Cardano, therefore, complex numbers seemed useless and sophistic. He never appeared happy and competent in their employment.[64]   On the other hand Bombelli, to whom we shall next turn, though starting off suspicious of these "sophistic numbers", nevertheless eventually managed to systematize them and certainly found them very useful and practical.

Chapter IV

Revelation

In this chapter we first see how Bombelli introduced complex numbers in a very pragmatic way but with quite remarkable insight. Then we look at how complex numbers were accepted over the next couple of centuries. Finally we see how Leibniz justified the rule for the solution of a cubic equation by means of complex numbers.

1. Bombelli

Cardano's obscure, or perhaps even obscurantist, treatment of complex numbers may explain why Bombelli can so positively claim that *he* has discovered a new kind of cube root, that is, a cube root involving $\sqrt{-1}$, when he writes:

I have found another sort of tied cube root $[\sqrt[3]{(a + \sqrt{b})}]$ very different from the others (Bombelli [1572], p.133).[1]

For Bombelli goes on, as we shall see, to a very thorough treatment of complex numbers as they arise in this context.

Bombelli was probably writing his *L'Algebra* between 1557 and 1560 (Jayawardene [1965], p.304), though he did not publish any of it until 1572. At that time only the first three (algebraic) books were published, leaving two more geometric books unpublished at his death in that same year. The complete work was finally edited and published by Bortolotti [1966].

Unlike the other sixteenth century Italians we have mentioned, Bombelli was not a university teacher but an engineer. It was during an interruption to work on the draining of the swamp of the Chiana in Italy that Bombelli began his *L'Algebra* (see Bortolotti [1903-4], p.61).[2] Having written the first draft of his book, he discovered a manuscript of Diophantos' work (preface to Bombelli [1572], p.8). Previously he had used the work of, amongst others, al-Khwārizmi, Fibonacci and Pacioli.

In writing his preface he says:

I began by reviewing the majority of those authors who have written on the question up to the present, in order to be able to serve instead of them on the matter, since there are a great many of them. Among them a certain Arab, Mohammed ibn Musa [al-Khwarizmi], is believed to have been the first; there exists a short work of his, of scant value, from which I believe this word 'algebra' has come down to us (Bombelli [1572], f. d2r).[3]

But of course Cardano's book (Cardano [1968] had appeared in 1545, so there was nothing startlingly new in the first part of Bombelli's work, which is basically a beautiful codification of existing knowledge.

The impact of the Diophantos manuscript was enough to make Bombelli rewrite a lot of his book. That there was a Diophantos manuscript had been known since before 1464: a manuscript is recorded in the Vatican library in 1455 (Rose [1976], p.37). It was Regiomontanus who drew attention to Diophantos. Writing in a preface to an edition of Alfraganus [= ibn-Kathîr] [1537], Regiomontanus says:

> But so far no one has put into Latin from Greek the thirteen very clever books of Diophantos, in which the very flower of all Arithmetic is hidden, the art of unknown and power which they call today by the Arab name: Algebra.[4,5]

Though Camerarius and Peletarius also remarked on finding a manuscript of Diophantos about a hundred years later (Diophantos [1890], p.v), it was Bombelli and a colleague, Pazzi, who first attempted a translation[6], probably in the 1540s. Bombelli writes in his introduction:

> But in recent years a Greek work in this discipline was rediscovered in the library of Nostro Signore in the Vatican, written by a certain Diophantos of Alexandria, a Greek author, who lived in the time of Antonius Pius, and it having been shown me by Antonio Maria Pazzi[7], Reggiano, professor of mathematics at Rome, and having judged, with him, the author to be well informed about number (although he did not treat irrational numbers, but in it there appears only a complete means of proceeding), he and I, in order to enrich the world with a work so finely made, decided to translate it and we have translated five of the books (there being seven[8] in all); the remainder we were not able to finish because of pressure of work on one or other. In the work [we have translated] we have discovered much that [the author] on several occasions quotes from Indian authors, from which I learnt that this discipline was known to the Indians before the Arabs (Bombelli [1572], p.8-9).[9]

Having found this manuscript he rewrote his earlier work so that it incorporated 143 of Diophantos' problems (Diophantos [1959], p.LXVI).

Bombelli starts off his book in a way very similar to Diophantos and throughout the work he has a very standardized approach. He succeeds in his aim of "reducing [algebra] to perfect order" (Bombelli [1572], p.8)[10]. He is exceedingly systematic. He uses notation similar to Chuquet and Pacioli.

It is clear from the development of polynomial equations with positive integral coefficients in Bombelli's book, where he gives an extensive treatment (Bombelli [1572], p.157 f.) of the manipulation of such expressions (their addition, multiplication and some divisions), that Bombelli is very much more in command of his algebra than Cardano was. Let us now consider his development in *L'Algebra*, for there lies the key to his accommodation of complex numbers.

Bombelli sometimes treats multiplication before addition. He progresses through three phases which correspond to numbers, complex numbers and polynomials. He starts with numbers and although his examples only involve integers, he gives the clear impression that this is not an essential restriction.   Each time he introduces a new kind of expression, for example a sum or product, he consistently gives quite explicit rules for their addition, multiplication and so on.   Thus Bombelli writes:

Multiplication of plus and minus

In order to make clearer the process of multiplication we give more examples.

> *Plus by plus makes plus.*
> *Minus by minus makes plus.*
> *Plus by minus makes minus.*
> *Minus by plus makes minus.*
> *Plus 8 by plus 8, makes plus 64.*
> *Minus 5 by minus 6 makes plus 30.*
> *Minus 4 by plus 5 makes minus 20.*
> *Plus 5 by minus 4 makes minus 20.   (Bombelli [1572], p.62)*[11]

And again, for a clearer understanding, there will be given further examples of composed numbers, such as binomia and residua [some numbers of the form a + √b] (ibid.).[12]

Then, after one example, (6+4) times (5+2), he gives the following:

To multiply (6+4) by (5-2).   Minus 2 by + 4 makes - 8, and - 2 by 6 makes - 12, because the 6 is plus, because there is no minus sign, and 5 by + 4 makes + 20, because the 5 is plus, and 5 by 6 makes 30, which is plus, because there is no minus sign: so altogether the products of the multiplication will be 30 + 20 - 12 - 8, but I shall not give here the proof of this multiplication otherwise, because we have not yet given rules for the sum of plus and minus (ibid.).[13]

$$\begin{array}{r} 6 + 4 \\ 5 - 2 \\ \hline 30 + 20 - 12 - 8 \end{array}$$

Thus Bombelli is explicitly working with signed numbers.   He has no reservations about doing this, even though in the problems he subsequently treats he neglects possible negative solutions.

Next he gives the rules for subtraction of signed numbers.   Then comes the definition of *binomium* and *residuum* (which are expressible[14] in the form a ± √b) with rules for operating with them.   Immediately following we come to

Demonstration of how minus times minus makes plus.[15]

The demonstration is purely geometrical:

Let the line gi be √18, from which we have to subtract the line m = √2; on the given line gi let there be marked the point h in such a way that gh is equal to the line m, and in order to know how much is the remainder hi construct above gi the square acgi and then from the point h draw hb

parallel to ag and in ag put the point d in such a way that dg is equal to gh and to that point d draw df parallel to gi and through the [...] of the second the parallelogram bcef will be a square, and will have a side equal to the line hi remaining from gi. And to find how big this square is, it is known from the given gi, which is √18, that the square acgi is 18 in area; but if from that one takes the gnomon bgf [that is the l-shaped figure adghefcba] there remains this square bcef. And to know how big the gnomon is: it is known that ab and gh are in parallel, and that since the line ag is √18 and the line ab is √2, multiplying one by the other yields √36, whose [square] root is 6, and thus the parallelogram dfgi is therefore 6, since it has sides of the same length. But to know how big the parallelogram efhi alone is one has to take away degh which is 2 because it has a side of length gh which is √2. Therefore, all the gnomon bgf is 10 which, subtracted from 18 leaves 8, and thus the line hi will be √8. (Bombelli [1572], p.77)[16]

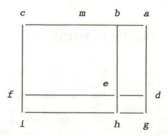

This type of geometrical demonstration is usual with Cardano too, as we have seen. We note that the geometric treatment guarantees the commutativity and associativity of addition, as well as agreeing with Bortolotti (see his note 30 on p.77 of Bombelli [1572]) that Bombelli here, and previously, was operating with the distributive laws.

In his approach to polynomials, Bombelli's treatment is very similar. First he describes how to multiply and divide powers of the unknown (though he does not consider negative exponents). Then come the operations with signed multiples of powers. He makes it clear that the powers added, etc., must have the same exponent. (He writes 3^1 for 3x, 5^2 for $5x^2$, and so on.)

Taking 3^1 from 8^1 will leave 5^1, but taking 3^1 from 5^2 cannot be written otherwise than $5^2 - 3^1$ (Bombelli [1572], p.161).[17]

Then come subtraction, multiplication and division for signed multiples of powers. These completed, he moves on to polynomials (*dignità composte*) treating addition before multiplication (*ibid.*, p.163 f.).

Before he deals with general polynomials he expands from dealing with *binomia* and *residua* to expressions involving cube roots and his bracketing system: he uses the symbols ⌞ and ⌟ in the printed version of his book for our (and), but these are joined in the manuscript.[18] This has the effect of allowing long sequences of symbols without intervening words, so that we read, for example, on p. 133,

16.p.R.c. ⌞R.q.5640192.p.2368⌟.m.R.c. ⌞R.q.5640192.m.2368⌟

or in modern notation

$$16 + \sqrt[3]{(\sqrt{(5640192)} + 2368)} - \sqrt[3]{(\sqrt{(5640192)} - 2368)}.$$

All these calculations and the facility which Bombelli appears to have developed in manipulating roots and tied roots in the style of Chuquet pave the way for his new discovery.

2. Bombelli and Complex Numbers

Before he introduces complex numbers explicitly, Bombelli points out that they arise in the solution of equations of the form $x^3 = ax + b$ when $(a/3)^3 > (b/2)^2$, that is, when the discriminant of the cubic is negative. Here the solution involves taking the cube root of a number plus the square root of a negative number. He calls this a "tied cube root" for the ligature (bracketing) ties the two parts, the number and the imaginary number, together.

He also calls $+ \sqrt{-}$ "plus of minus" and $- \sqrt{-}$ "minus of minus". Although this might appear to be an abbreviation of "plus square root of minus" as in $+ \sqrt{-a}$, Bombelli in fact seems to use "plus of minus" in the way we use i, the square root of minus one, thus $+\sqrt{-a}$ means $i\sqrt{a}$.

Bombelli writes:

I have found another kind of tied cube root very different from the others, which arises from the case of the cube equal to the unknown and a number [i.e. the solution of the cubic equation $x^3 = ax + b$] when the cube of one third of [the coefficient of] the unknown is greater than the square of half the number [i.e. $(a/3)^3 > (b/2)^2$] as will be shown in that chapter. This sort of square root has for its algorism[19] different operations from the other and a different name; because when the cube of one third of [the coefficient of] the unknown is greater than the square of half the number, the excess cannot be called either positive or negative, but I shall call it plus of minus [piu di meno] when it has to be added and when it has to be taken away I shall call it minus of minus [men di meno], and this operation is very necessary, more than the other tied cube root for the matter of the fourth power with a cube or an unknown or with both together [i.e. for quartic equations], where there are many more cases of equations producing this kind of root than of those producing the other kind. This

Having dealt with the multiplication of real and purely imaginary quantities, he goes on to the multiplication of numbers formed using square and cube roots as well as addition and subtraction. For example, $\sqrt[3]{}(2+i) + \sqrt[3]{}(2-i)$ times 4 (*ibid.*, p.134). Sometimes he multiplies complex conjugates, for example $\sqrt[3]{}(3+i\sqrt{}2)$ times $\sqrt[3]{}(3-i\sqrt{}2)$ (*ibid.*), but although complex conjugates feature a great deal in this part of his work, he considers products of other complex numbers. Thus we find, for example, $\sqrt[3]{}(4+i\sqrt{}2)$ times $\sqrt[3]{}(3+i\sqrt{}8)$ (*ibid.*, p.135).

It is only after all these computations that he turns to sums and differences of complex numbers. First recall our remarks above, that "plus of minus" a means ia where i is the square root of minus one. Bombelli goes on to explain how to add ia to –ib etc..

Summation *of + of - and - of -.*

The summation of + of - and - of - has the rule (as for the other [numbers]) which we shall present with the usual brevity. Plus cannot be added to + of -, one does not say plus + of -, as if one were to say "add + 5 to + of - 8; that makes 5 + of - 8", and the same goes for - of -. Plus of - with + of - sums and makes + of -. Plus of - with - of - subtracts and the result has the sign of the larger quantity [in modulus]. Minus of - with - of - sums and makes - of -. Sum + of - 8 and - of - 5: that makes + of - 3. Sum + of - 15 and - of - 28: that makes - of - 13. Sum - of - 12 and - of - 6: that makes - of - 18. Sum + of - 6 and + of - 15: that makes + of -21. And, the operation being clear from these examples above, I shall show under tied cube roots what is importance is and where the case can arise (ibid., p.147).[26,27]

Then comes "Summation of tied cube roots of + of – and – of –."[28] followed by subtraction of + of – and – of – and subtraction of cube roots involving imaginaries (*ibid.*, p.148-149).

Thus we have an engineer, Bombelli, making practical use of complex numbers, perhaps because they gave him useful results, while Cardano found the square roots of negative quantities useless. Bombelli is the first to give a treatment of any complex numbers and even in his earlier draft he used complex numbers (Bombelli [1572], p.8-9 and Bortolloti [1929], p.417). It is remarkable how thorough he is in his presentation of the laws of calculation for complex numbers, for it appears that at first he merely followed the formula for solving a cubic blindly and then found that it worked in *all* cases.

At that time there was no question of proving that the formula always worked, for an acceptable proof would have had to have been presented in the Euclidean geometric style: the style which Cardano used in his *Ars Magna* (see above, Chapter III, §9). But when Cardano discussed quadratics with imaginary roots, he left it to the reader to imagine $\sqrt{}-15$ and Bombelli did not even attempt to give any geometric representation of any complex number in the algebraic books of his work.

Before we leave Bombelli, we should mention that he also used geometric methods for solving equations. In addition to ruler and compass he used a carpenter's square, that is to say, a fixed right angle with arbitrarily long sides. This enabled him to give demonstrations for his solution of equations in the plane instead of in (three-dimensional) space in his *L'Algebra*.

Now, although Bombelli claimed that he had proved that his computations with complex numbers gave genuine roots of cubics using Scipione's formula, it was not until more than a century later that Leibniz gave a general, algebraic proof (see below, §7). In the intervening period, however, many authors worked with, dallied with or derided the new kind of number. We now turn to some of these.

3. Stevin

Simon Stevin, whose *L'Arithmétique* appeared in 1585, 13 years after Bombelli's *L'Algebra*, was not prepared to use Bombelli's new numbers. As we shall see below, he did, however, passionately assert the view that all numbers are of one kind:

> THAT THERE ARE NO ABSURD, IRRATIONAL,
> IRREGULAR, INEXPLICABLE OR SURD NUMBERS.

It is a very common thing amongst authors of arithmetics to treat numbers like $\sqrt{8}$ and similar ones, which they call absurd, irrational, irregular, inexplicable, surds &c. and which we deny to be the case for any number which turns up: But by what reason will the adversary prove it unreasonable? (Stevin [1958], vol.II, p.532.)[29]

Zero and negative numbers, nevertheless, are somewhat difficult to handle and he will not use Bombelli's new numbers:

> As for me, I regard it as useless to write here about such; The reason is, that what cannot be found by definite rule seems unworthy of a place amongst legitimate propositions. On the other hand, for that which is solved in such a fashion Fortune deserves as much honour as the operator [i.e. user of the method]. Thirdly, that there is enough legitimate matter, even infinitely much, to exercise oneself without occupying oneself and wasting time on uncertainties: therefore we shall leave these aside. Those who take pleasure in such examples can do with them as they choose (ibid., p.619-620).[30]

Stevin does not treat equations as such. Like Viète (see below, §4), he discusses problems about proportionals. Finding proportionals is equivalent to solving polynomial equations. For the case of the cubic compare §4 below. Stevin uses a power notation akin to Bombelli's. So he writes (*ibid.*, p.672):

$$1 \; ④ \; - \; 4 \; ③ \; + \; 12 \; ② \; - \; 16 \; ① \; + \; 16$$

where we would write $x^4 - 4x^3 + 12x^2 - 16x + 16$ and sometimes he uses ⓪ where Cardano writes N to stand for an arbitrary (natural) number. We shall return to the fact that Stevin uses negative coefficients below.

In his basic treatment of the arithmetic operations he has proceeded, like Bombelli, to give a theorem:

Plus multiplied by plus, gives product plus, and minus multiplied by minus, gives product plus, and plus multiplied by minus, or minus multiplied by plus, gives product minus (ibid., p.560).[31]

Stevin claims to give this theorem an arithmetic demonstration and also a geometric one. In fact, both demonstrations actually treat (8 - 5) times (9 - 7) equals 6 and the geometric one does not contain negative lengths or areas. Neither does Stevin treat of sums involving negative numbers.

Further, very early in his book he only distinguishes two kinds of number: geometric and arithmetic.

The one [kind] is explained using adjectives of size, like square numbers, cubic [numbers], roots, quantities, &c; these we call Geometric numbers, and they will be defined in the second part below; The other kind is simply explained without any adjective, like one, two, three fifths, &c. (ibid., p.504).[32]

He is then often not constrained by considerations of (geometric) homogeneity of dimension, unlike Viète, who in his *Isagoge* (1596) pronounces:

The first and permanent law of equalities or proportions which, because it is conceived from homogeneous [quantities] is called the law of homogeneous [quantities], is this:

Homogeneous [quantities] must be compared with homogeneous [quantities] (Viète [1646], p.2.).[33]

Stevin classifies the problems in terms of the rule of three.[34] He acknowledges the work of al-Khwārizmi, Scipione dal Ferro, Tartaglia, Cardano and Diophantos, though he appears to think that Diophantos followed al-Khwārizmi.

As for Diophantos it seems that in his time the discoveries of Mahomet [al-Khwārizmi] had only been known as could be found in the first six books; ... (Stevin [1958], p.585).[35]

As we have just seen, Stevin essentially follows Bombelli's exponential notation, but when it comes to negative coefficients he has no hesitation at all. Indeed, his first example (*ibid.*, p.570) is to multiply 2③ - 4② + 3① by 2④ + 3③ (In modern notation: $2x^3 - 4x^2 + 3x$ by $2x^4 + 3x^2$.) Stevin's treatment of such polynomials

97

extends to long division and finding the greatest common divisor (*trouver leur plus grande commune mesure*) (*ibid.*, p.477).

Subsequently he proceeds to finding fourth proportionals (*ibid.*, p.593 f.). This includes equivalents of solving quadratic and cubic equations. Stevin provides geometric demonstrations in the style of Cardano and he uses Cardano's rule for cubics (*ibid.*, p.612 f.). When he comes to the irreducible case, he writes:

ON THE IMPERFECTION THAT THERE IS IN THIS FIRST DIFFERENCE

Rafael Bombelli solves it by the locution of plus of minus & minus of minus (ibid., p.618).[36]

He gives an example with solution $\sqrt[3]{18 + \text{of} - 26} + \sqrt[3]{18 - \text{of} - 26}$. He then notes that

if by the numbers of this solution one knew how to approach infinitely [close] to 6 (for they are precisely of that value) as one does by the numbers of the solution of the foregoing first example, certainly this difference would be of the desired perfection (ibid., p.619);[37]

but he then writes the disclaimer noted above (our section quotation in this section).

Stevin's treatment of negative roots is novel.

Some of the preceding problems, on the proportion of algebraic numbers, get, in addition to the solutions given by the above, still another solution by - ; And although the same seem only fanciful solutions, all the same they are useful, as a means of arriving at the true solutions of the following problems by + ; The reason is, that to the value of [x] found by some of the preceding problems, it will be necessary sometimes still to add some certain number, as it will appear; whence it follows that, when the number to be added will be greater than the solution by -, that their difference will be a true solution by + . (ibid., p.642).[38]

In other words, he transforms an equation such as $x^2 + 5x + 6 = 0$ which only has negative roots -2 and -3 by replacing, say, $x + 4$ by y, thus getting $y^2 - 3y + 2 = 0$ which has positive roots 1 and 2 for y which are "true solutions" but which nevertheless give the roots -3 and -2 of the original equation. (But he does not even mention the possibility of zero being a solution of an equation.) On the other hand, if he obtains a positive solution then he does not mention negative solutions at all. He, therefore, would have different numbers of roots for different cubic equations (*cf. ibid.*, p.648).

4. Viète

Stevin's slightly older contemporary Viète (1540-1603) did not publish his work until 1591 but his work, while accepting negative coefficients, is constrained by the law of homogeneity mentioned above. Viète uses letters A, B, C, etc. to denote arbitrary magnitudes. This permits him to give general solutions and also to state identities. Like Stevin, Viète makes it explicit that proportionals give solutions of equations.[39] Indeed, the two ways in which Viète solves equations are: 1. by reducing them to problems of finding proportionals, and 2. by comparing them with known, solved equations.[40] Thus in considering $A^3 + B^2A = B^2Z$ where A is unknown, Viète writes:

If $A^3 + B^2A = B^2Z$: there are four continuing proportionals, of which the first greater or less between the extremes is B, but the sum of the second and the fourth is Z and A is the second. The second being A, the fourth will therefore be Z - A. But the solid [which is the product] of the first squared and the fourth is equal to the cube of the second: since as the square of the first is to the square of the second so is the second to the fourth. Therefore $A^3 = B^2Z - B^2A$; and by transposition $A^3 + B^2A$ will equal B^2Z as is said (Viète [1646], p.86).[41]

This is the first theorem related to cubic equations which Viète gives.[42] He then goes on to present the other theorems according to his classification treating $A^3 - B^2A = B^2D$, $B^2A - A^3 = B^2D$, etc. For Viète, however, solutions are only part of the story. He spends considerable space on theorems, such as "If $A^2 + B = Z$ then $A^4 + (B^5 + 2BZ)A = Z^2 + B^2Z$" (where Z is a plane number to get the dimensions right). What the status of such theorems and their relation to the solution of problems is, is perhaps indicated near the end of *De Emendatione Tractatus Secundus*, where Chapter IX bears the legend:

Anomalous reduction of some cubic equations to quadratic or even simpler ones (Viète [1646], p.152).[43]

Viète is systematic in his use of letters for both coefficients and unknowns. Then by comparing the forms of equations he can give special cases as examples. Thus we find not only "If $A^3 - 12A = 16$ then $A = 4$", but also the preceding example "If $18A - A^3 = 27$ then $A^2 + 3A = 9$". In this way he progresses to a final chapter (Viète [1646], p.158) where by comparison with an equation such as

$$A^3 + (- B - D - G)A^2 + (BD + BG + DG)A = BDG$$

he obtains solutions $1, 2, 3$ for $A^3 - 6A^2 + 11A = 6$. Here at last we are beginning to see an abstract study of polynomial equations.

5. Harriot

Between Viète and Descartes we find the English mathematician Harriot. Stevens wrote:

From [the introduction to Harriot's **Artis Analyticae Praxis** *(1931) by Aylesbury and Prothero assisted by Warner] it appears that Hariot's system of Analytics or Algebra was based on that of his friend and correspondent François Vieta as Vieta's was avowedly based on that of the ancients ... Full credit was given by Hariot and his friends to the distinguished French mathematician (Stevens [1900], p.153).*

Thomas Harriot (1560-1621) did not publish any mathematics in his lifetime and the *Artis Analyticae Praxis* (Harriot [1631] appeared ten years after his death, edited by others. Unfortunately it is generally agreed[44] that it was very badly edited.

Harriot's work is stark. Although it begins with wordy definitions it then proceeds to use algebraic symbolism which could have been written yesterday with only the slightest change (see for example Rigaud [1832] where a page of Harriot's manuscript is reproduced). It is almost exclusively symbolic; numerical examples are not treated till the end (Harriot [1631], p.117).

Wallis used a lot of Harriot's material in his *Algebra* (Wallis [1684/5].[45] There we find a thorough-going and highly developed treatment of quadratic and cubic equations and complex roots.

Thus in Chapter XXXII, Wallis says Harriot showed there are always two roots to a quadratic equation either real or imaginary (that is, complex). Wallis quotes the example: $-aa + 8a = 25$ whence $a = 4\pm\sqrt{:16-25}$ so that $a = 4 + \sqrt{-9}$ or $4 - \sqrt{-9}$ which are *radices imaginariae* (imaginary roots). He then continues:

But howsoever they be Imaginary, or Impossible, equations of this type are nevertheless not useless: but they have uses not to be despised, as will be explained in its place (ibid.).[46]

Harriot derived cubics from considering $a = +b$, $a = +c$, $a = +d$ and then taking the product $(a-b)(a-c)(a-d) = 0$. This is very close to Viète's work quoted above at the end of §4 (Viète [1646], p.158).

In Harriot's solution of cubic equations (as found in Wallis [1684/5], p.181 f.), we find an almost entirely algebraic presentation of what is now known as Viète's method. Paraphrasing we find:

Consider $a^3 + 3b^2a = 2c^3$

Having put $a = \dfrac{e^2-b^2}{e}$ we get

$$e^6 - 2c^3e^3 = b^6$$

and hence

$$e = \sqrt[3]{(c^3 + \sqrt{(b^6+c^6)})}.$$

Then to find a Wallis (following Harriot) says:[47]

Note e^3, b^3, $b^6/3$ *are continually proportional quantities. And so too are their cube roots* e, b, b^2e

whence $b^2/2 = \sqrt[3]{(-c^3+\sqrt{(b^6+c^6)})}$. For multiplying $c^3 + \sqrt{(b^6+c^6)}$ by $-c^3 +\sqrt{(b^6+c^6)}$ we get b^6.

Also, when the former equals e^3, *the latter will be equal to* b^6/e^3, *and* b^3 *is the middle proportional between those; and its cube root will be equal to* b^2/e.

Hence the value of $a = \sqrt[3]{(c^3 + \sqrt{(b^6+c^6)})} - \sqrt[3]{(-c^3+\sqrt{b^6+c^6})}$. Wallis adds:

... these are clearly derived by Harriot's method, and thence demonstrated; very different from Cardano's method (ibid., p.183).[48]

This is indeed Harriot's method and occurs on p.98-99 of Harriot's [1631]. Wallis has followed Harriot to the letter.

In the irreducible cubic where one has to take square roots of negative numbers, that is, in $a^3 - 3b^2a = +2c^3$ where $0 < c < b$, Wallis says Harriot was the first to show (*demonstrare*) that the equation had a real positive root (Radicem habere *Affirmativam Realem*) and also two negative roots (Wallis [1684/5], p.183).

Although he gives numerical examples, Wallis points out (*ibid.*, p.186) that he knows no general method for eliminating the imaginary quantities and the method (*ibid.*, ch. XLVI, p.185) that he gives follows Cardano's geometric demonstration (see above, chapter III, §3).

No-one before Leibniz appears to have done a complete algebraic check that Cardano's solution satisfies the cubic equation, though Harriot comes very close to doing one. Wallis, again mentioning Harriot (*ibid.*, p.192), considers the equation $r^3 - 63r = 162$. He says that by the rule $r = \sqrt[3]{(81+30\sqrt{-3})} + \sqrt[3]{(81-30\sqrt{-3})}$. He now assumes the cube roots to be of the form $a \pm f\sqrt{-e}$ and proceeds to compute their cubes as

$$a^3 + 3a^2f\sqrt{-e} - 3af^2e - f^3\sqrt{-e}$$

and

$$a^3 - 3a^2f\sqrt{-e} - 3af^2e + f^3e\sqrt{-e}.$$

Then putting $\sqrt{-e} = \sqrt{-3}$ and $f = 1$ he checks that he has indeed a solution of the cubic (though he does not say whence he obtained the $\sqrt{-3}$ and the 1). Finally the root r is

$$(a + f \sqrt{-e}) + (a - f \sqrt{-e}) = 9,$$

the imaginary parts cancelling.

Wallis completes the solution by computing the remaining roots using long division and then solving the resulting quadratic.

It is thus clear that Harriot had advanced beyond Viète in algebraic manipulative facility and in the matter of how many roots a quartic, for example, had.[49]

Finally, it is clear that Harriot accepted complex numbers as solutions of equations without hesitation.

6. Descartes

Descartes, who followed Harriot, was adept in the use of symbols. He appears to have regarded them as aids to geometry more than as valuable in their own right.

Often there is no need to draw the lines on paper, and it suffices to designate them by letters, each [line] by a single letter (Descartes [1954], p.298-299).[50,51]

Descartes was still subject to a version of the law of homogeneity (of dimension):

Here it must be observed that by a^2, b^3, *and similar expressions, I ordinarily mean only simple lines, which, however, I name squares, cubes, etc., so that I may make use of the terms employed in algebra. It should also be noted that all parts of a single line will normally each be expressed by as many dimensions as any other, whenever the unit is not defined in the question ... (ibid., p.229).*[52,53]

Like Newton (see chapter III, §6 above), Descartes classified equations in terms of the curves required to give a solution (*ibid.*, p.315). Descartes, in writing equations, had no inhibitions about using negative coefficients, but he was ill at ease with negative roots. Thus, when faced with a negative solution, he (essentially) changes the appropriate signs in a way similar to Stevin (*ibid.*, p.326). Again we read:

It is true that if the point E is not on the same side of the curve as C, only one of these will be a true root, the other being reversed, or less than nothing [i.e. negative] (ibid., p.346-347).[54]

He did, however, recognize the possibility of repeated roots, for he then goes on:

*The nearer together the points C and E are, however, the less
difference there is between the roots; and when the points coincide, the
roots are exactly equal, that is to say, if the circle through C touches
the curve CE there [at C] without cutting it (ibid.).*[55]

Cubic equations, and indeed equations of higher degree, arise for
Descartes from the search for proportionals (*ibid.*, p.370 f.).[56]   Having
shown that equations arise thus, he goes on to treat equations in general,
noting that false (i.e. negative) roots become true (i.e. positive) and *vice
versa* by changing the signs of the coefficients of the odd powers of the
unknown in an equation.

Complex roots are briefly considered but not used:

*Moreover, both the true and the false roots are not always real, but
are sometimes only imaginary, that is, one can always conceive of as many
[roots] for each equation as I have stated; but sometimes there is no
quantity corresponding to those [roots] conceived of (ibid., p.380).*[57]

*Now these two equations have no roots either true or false, whence we
know that the four roots of the original equation are imaginary; ... (ibid.,
p.386).*[58,59]

The last passage continues:

*... and that the problem whose solution depends upon this equation is
plane, but that its construction is impossible, because the given quantities
cannot be united (ibid., p.386).*[60,61]

This passage points up Descartes's overriding concern with geometry.
For Descartes, algebra is always only an aid to solving geometric problems.
He is quite prepared to do complicated algebraic manipulations in the course
of solving a problem, but he often reverts to geometric methods to solve an
equation.   Thus when an equation of degree three or four is to be solved,
he transforms the problem into a geometric one and then gives a construction
(*ibid.*, p.390 f.).   It is inevitable that in this way he should avoid
complex numbers as they cannot be represented in his geometrical setting.

Descartes discusses Cardano's rule (*ibid.*, p.390 f.) pointing out the
difficulty in the irreducible case where square roots of negative quantities
are involved.   But now, when dealing with the equation
$z^3 = pz + q$ (p,q > 0) where $(q/2)^2 \leq (p/3)^3$, that is, in the irreducible
case, Descartes returns to Viète's method of finding mean proportionals. On
the other hand, for the same equation when $(q/2)^2 > (p/3)^3$, he uses
Cardano's solution – but grudgingly:

*Moreover, it is to be noted that this mode of expressing the value of
the roots by the ratio they have to the sides of certain cubes of which one
only knows the volume, is no more intelligible nor simple than expressing
them by the ratio they have to chords of certain arcs or parts of circles
whose triple is given.   So much so that all the cubic equations which
cannot be expressed using Cardan's rules can be expressed at least as
clearly or more clearly by the method proposed here (ibid. p.400).*[62]

7. Leibniz

Now, although Bombelli claimed that he had proved his computations with complex numbers gave genuine roots of cubics using Scipione Dal Ferro's formula, his actual *geometric* demonstration does not represent complex numbers at all. This was pointed out in a very perceptive letter from Leibniz to Huygens, probably written about March 1673 (Leibniz [1899], p.548). Before considering Leibniz's remarks, we should bear in mind that the development of algebraic symbolism by Viète and its systematic exploitation by Descartes[65] had led to the consideration of cubic equations as general as any we consider today. There was a slight difference in emphasis as coefficients were expected to be rational, though other possibilities (for example, surds) were contemplated (Leibniz [1899], p.548). Moreover, negative roots, called *false* by Cardano and Bombelli[64], were accepted in Descartes's time.

We turn to Leibniz's letter:

I am sending you Bombelli's book, of which I spoke to you. You will see there, (Bombelli [1572], p.225] how he uses imaginary roots (for example, he calls $\sqrt{-121}$ or $11\sqrt{-1}$ plus of minus 11; and $-\sqrt{-121}$ or $-11\sqrt{-1}$ minus of minus 11) and how he finds the root of the equation[65] $1^3 = 15^1$ plus 4, that is, $y^3 = 15y + 4$ by means of that. He says he has a geometric proof, which he also inserts [ibid., p.231], but there he only proves that such an equation is possible, and that its root is something real which can be given geometrically. But it does not follow that the operation using his plus of minus is sound. For although he says at the end of it [ibid., p.226] that these roots come from the equation, this is not, however, without any assumption. It also appears by that [ibid., p.226], that he was not able to solve by this method $y^3 = 12y + 9$ whose rational root is false or negative, that is to say -3. Nevertheless in trying another method (also taken from Cardano) he finds that the equation is divisible by $y + 3$, not knowing that for that very reason -3 is the false root: ... and he finds by this [other] means the true [root] $1\frac{1}{2} + \sqrt{5\frac{1}{4}}$, which, being composed of a [rational] number and a square root, cannot be obtained from Cardano's formulas, because the roots one gets from these formulas, are always either irrational cube [roots] or [rational] numbers. Because of this he believed that Cardano's formulas did not work in this event, and are not general (Leibniz [1899], p.527).[66]

First Leibniz points out that finding a (correct) root of a cubic equation by complex numbers does not justify complex numbers. "But it does not follow that the operation using his plus of minus is sound." Then he points out that when Cardano's formula gives a real root, it involves either cube roots or rational numbers and never square roots (when the coefficients are rational). This we explain today by saying that in the case of three real roots, the galois group is cyclic of order three and when there is one real root we get a pair of complex conjugate roots and a galois group of order two.[67]   How Leibniz came to his conclusion is not known.   Since, in the later part of this letter, he claims to have a general method for solving equations even of the fifth degree[68], it seems reasonable to conclude that he was working from a combination of practical experience and intuition rather than being the possessor of a sound mathematical proof of his assertion.

His letter continues:

So I believe I have been the first to demonstrate (1) that Cardano's formulae are absolutely valid and general, whether [the roots] be extractable or non-extractable, whether true or false or negative. (2) That by this means we have the general solution of all cubic equations. (3) I have been the first to find that one can form complex non-extractable roots of all the even degrees which contain imaginaries and yet whose reality can be rendered palpable without extraction, in order to demonstrate that the reality of such formulae is not limited by extractability: of this the example of the formula $\sqrt{1+\sqrt{-3}} + \sqrt{1-\sqrt{-3}}$, which has value $\sqrt{6}$, is quite a considerable proof.   (4) I show, what no-one has shown before, that every cubic equation, which can be depressed [in degree], contains a rational root, provided the equation itself be given in rational terms.   From which it follows that [an equation] which cannot be divided by the unknown + or - a rational divisor of the last term, is solid [i.e. indecomposable].   [This is] a very important proposition, since it gives us a sure means of knowing if a problem is solid in fact, or only in appearance.   M. Descartes does not speak so positively, for he says, that one must examine all the quantities which may divide the last [term], which he supposes to be an integer or rational: and it seems that he dare not say all numbers, or all rational quantities.   So that he leaves us in doubt whether one must also examine the irrational divisors: whether this be because he had no sufficiently convincing demonstration for rational divisors to the exclusion of irrationals, or whether he did not bother to speak more precisely.   From which one can also show, in the fifth place (5) by analysis alone, without the aid of Geometry that every cubic equation is possible [to be solved], provided it is expressed in possible terms (Leibniz [1899], p.547-548).[69]

Leibniz did not publish the details of these claims but Gerhardt includes in the *Briefwechsel* a polished working paper of Leibniz which contains some proofs.

After a discussion of complex numbers obtained from quadratic equations and subsequent manipulations of expressions involving cube roots which, though very interesting[70], are not relevant here, he shows that the equation

$$x^3 - 48x - 72 = 0$$

has a root $x = -6$ and that the expression given for the root by Cardano's expression, namely

$$\sqrt[3]{(36 + \sqrt{(-2800)})} + \sqrt[3]{(36 - \sqrt{(-2800)})}$$

can be reduced, by algebraic manipulation, to -6, too.

Having done this, he proceeds to the general case:

However so that no cause for doubt be left, let us give a double general proof which is not hindered by rationals or irrationals.   The former demonstration comes back to this: Cardano's expression satisfies the cubic equation with three [real] roots; every expression which satisfies some equation, is a root of it; therefore Cardano's expression is a root of the cubic equation with three [real] roots.   Again every root of the cubic equation with three [real] roots is a real quantity (indeed, for that very reason — because it has three real roots — we call it [an equation with] three [real] roots; exactly what it was that was believed not to fit Cardano's rules has already been shown; see first Schooten's appendix on the solution of cubic equations; and that in fact any cubic cannot have more than three).   Therefore Cardano's expression (even when it is derived from a cubic with three roots) is a real quantity.   Therefore it only remains for us to show that Cardano's expression satisfies even the cubic equation with three [real] roots; this is clear, if in an equation of this kind, such as $x^3 - qx - r = 0$, substituting the value of the x, in fact [if we write $x = A + B$ where A and B are the two cube roots then we get for $(A+B)^3$]

$$\begin{array}{cccc} A^3 & + 3A^2B & + 3AB^2 & + B^3 \end{array}$$

$$x^3 \sqcap \frac{r}{2} + \sqrt{\frac{r^2}{4} - \frac{q^3}{27}} + \sqrt[q]{(3)\frac{r}{2} + \sqrt{\frac{r^2}{4} - \frac{q^3}{27}}} + \sqrt[q]{(3)\frac{r}{2} - \sqrt{\frac{r^2}{4} - \frac{q^3}{27}}} + \frac{r}{2} - \sqrt{\frac{r^2}{4} - \frac{q^3}{27}}$$

$$- qx \sqcap - \sqrt[q]{(3)\frac{r}{2} + \sqrt{\frac{r^2}{4} - \frac{q^3}{27}}} - \sqrt[q]{(3)\frac{r}{2} - \sqrt{\frac{r^2}{4} - \frac{q^3}{27}}}$$

$$- r \sqcap - \frac{r}{2} \qquad\qquad\qquad - \frac{r}{2}$$

$$\sqcap 0 \qquad \sqcap 0$$

Therefore Cardano's expression always satisfies [the cubic] and it does not matter whether $q^3/27$ is less than $r^2/4$ (Leibniz [1899], p.560).[71,72,73,74]

That is, Leibniz has taken the algebraic expression for x given by Cardano's rule in the general case, i.e.

$$x = \sqrt[3]{\frac{r}{2} + \sqrt{\frac{r^2}{4} - \frac{r^3}{27}}} + \sqrt[3]{\frac{r}{2} - \sqrt{\frac{r^2}{4} - \frac{q^3}{27}}} = A + B.$$

Then, using $(A + B)^3 = A^3 + 3A^2B + 3AB^2 + B^3$ he has formally calculated x^3. (Note that he writes $\sqrt{(3)}$ for $\sqrt[3]{}$ and $\sqcap$ for $=$.) Using the fact that $AB = q/3$ "as is easily shown", he then sums x^3, $-qx$ and $-r$ and finds all terms cancel quite formally.

Thus Leibniz has shown that by formally manipulating expressions involving square roots of negative numbers in the same way as numbers of algebraic expressions, Cardano's formula always gives a root of a given cubic equation.

One of the most striking of his calculations with imaginaries, however, was very simple to state. He showed that $\sqrt{(1+\sqrt{-3})} + \sqrt{(1-\sqrt{-3})} = \sqrt{6}$.

Huygens was impressed:

The remark you make concerning inextractable roots, and with imaginary quantities, which however when added together yield a real quantity, is surprising and entirely novel. One would never have believed that $\sqrt{1 + \sqrt{-3}} + \sqrt{1 - \sqrt{-3}}$ make $\sqrt{6}$ and there is something hidden therein which is incomprehensible to me (ibid., p.566).[75]

8. Conclusion

The geometric aspect of equations prevented the development of solutions other than positive ones for a long time. Diophantos started to treat powers higher than cubes, thus paving the way to the algebra which was introduced by the Arabs about the end of the first millennium A.D. Mohammed al-Khwārizmi solved all "reasonable" quadratics and Omar Khayyam solved cubics by invoking conics.

It was not, however, until the sixteenth century that Scipione dal Ferro's solution of the cubic, by a formula, opened the way to the algebraic solution of cubic and quartic equations. Even then the formula had a geometric demonstration and it was Bombelli's insistence on applying the formula in cases where it seemed inadmissible, that led him to find a new kind of number which obeyed somewhat different rules. He manipulated these complex numbers, but did not give algebraic (or geometric) proofs justifying his use of complex numbers. He only checked that his answers were correct. A century and a half later, Leibniz at last gave a purely algebraic justification of Scipione's formula for the general cubic equation.

It had long been known that every number could be squared. The move that led to making sense of the inverse operation was long delayed. What finally happened was that Bombelli simply wrote down the formal application of the converse operation without regard to any interpretation. Nevertheless, carrying out formal calculations led him to answers which were correct. He had no justification of the process; he only had confirmation of the correctness of his results by substituting his answers back in the original problem.　　It was not until Leibniz's work over a hundred years later that a formal justification was provided.

It is important to note that this justification comes from the *acceptance* of the manipulation of symbols in a way far beyond that of the manipulation of equations produced by the Arabs.

Thus the introduction of complex numbers was accompanied by grave misgivings (on the part of Cardano in particular), accepted because it gave correct results (by Bombelli) and only subsequently justified (by Leibniz in particular) when the concept of algebra and algebraic manipulations had undergone significant further development.　　The first employment of complex numbers was a move into the unknown using known techniques.　　Its justification was initially pragmatic.

Here we have a stark contrast with the natural numbers, for they came into being very slowly, while complex numbers erupted quite suddenly. Although one might now say they were always hidden under the surface of polynomial equations, they were not brought out until the first half of the sixteenth century.　　At that stage they emerged with remarkable clarity.

In the next part we shall see a different emergence again.

Part 3

Real Numbers

It is a relatively small step to go from the natural numbers to rational numbers, that is, fractions of the form p/q. All that is required is to think in terms of new units which are q times as big as before. Then measuring (the old) p by the new units gives p/q in the new scale. This is what is to be found in Euclid (see Book VII). Euclid writes of one number measuring another, by which he means that one is an integral multiple of the other (Euclid VII.5). However, the Greeks discovered that although one could measure one length or magnitude by another, there were also easily constructed lengths, e.g. the diagonal of the unit square, which could not be measured exactly. The diagonal of a square is *incommensurable* with the side of the square. They discovered irrational numbers. We shall study the background to this discovery in chapter V.[1]

Now the Greeks were quite clear that no infinite processes were involved in showing that certain numbers are irrational. In the nineteenth century, people began to think about *all* the points on the line and this notion of *all* necessarily involved consideration of the infinite. In chapter VI we shall see how this was treated. We shall also point out that a new axiom, which is mentioned very infrequently these days, is required. This axiom allows us to identify the geometric line and the number line; an identification which the Greeks never made. Finally we shall see how the formalization of our intuition of the line has developed quite recently and see that the way is open for further developments.

Chapter V

Irrationals

1. Introduction

The discovery of irrational numbers by the Greeks has long, though perhaps erroneously, been regarded as causing a great crisis in Greek mathematics. Thus in the first scholium to Book X of Euclid's *Elements* (see Heiberg's edition, Euclid [1883-1916], vol. V, p.417) we read (in translation):

The Pythagoreans were the first to make inquiry into commensurability, having first discovered it as a result of their observation of numbers; for though the unit is a common measure of all numbers they could not find a common measure of all magnitudes. ... there is a story of the Pythagoreans that the man who was the first to bring examination of these matters out into the open suffered shipwreck, and perhaps they were hinting at the fact that everything that is irrational in every case is accustomed to be hidden (sc. so that it remains) irrational and formless, and if anyone (ψυχή) were to make an assault on such a form of existence as this, he would simply make obvious and clear the fact that he is being carried under into the sea of creation and is being overwhelmed by the unstable surges of it. Such awe did these men have for examination of the irrational.

These remarks, written later than the time of Theon of Alexandria (end of the fourth century A.D.; see Heath's Euclid [1925], vol. I, p.66 and Kline [1972], p.25) form perhaps the most explicit description of the effect of the discovery of irrational (or incommensurable) numbers that we possess.

In this chapter we shall consider the mathematical background against which this discovery took place after looking at the Pythagoreans and their mathematics. Then we shall report on some earlier suggestions as to how irrational numbers were first produced and present some of our own. In doing this we wish on the one hand to stress the difficulty of trying to read the minds of individuals of whom we have virtually no record and on the other to attempt to show one possible coherent picture. We close this chapter by pointing out that, although we think of irrational numbers as infinite decimals, the Greeks avoided the infinite in their treatment of irrational numbers. In particular they scrupulously avoided identifying magnitudes and numbers.

In the next chapter we shall consider the subsequent developments which led to the introduction of the infinite. Our basic conclusion there will be that, after the connexion between (natural[1]) numbers and measurement of, for example, the length of the side of a field or of a geometrical figure had been made[2], the direction was reversed and a connexion sought in the opposite direction - that is to say, from lengths to numbers. To us in the twentieth century it seems natural to make both these connexions.

However, the logical necessity of the second connexion depends on the philosophical view, and in particular the view of the philosophy of mathematics, which is held.

2. Sources

Before we turn to the starting point for our treatment, Pythagoras, it is worth pointing out that in the East there was never any difficulty about irrational numbers. Thus, in India, irrational numbers such as $\sqrt{2}$ and $\sqrt{3}$ were used in the context of preparing altars for yajñas. As in Greece, these arose in the context of what is generally known as Pythagoras' theorem, though this result is to be found in *Sūtra* 1.37 of the *Bandhayana* of about 1000 B.C. in India. These numbers were simply used for geometric constructions.

In China, where mathematics was almost always conducted in terms of algorithms rather than proofs, there was no difficulty either. The Chinese simply calculated lengths, for example, to the degree of precision necessary. They, too, were familiar with the result which is Pythagoras' theorem. This result appears in the *Zhoubi suanjing* which dates from about 100 B.C., though it is believed its contents were known much earlier and did not derive from the Greeks (see Li Yan and Du Shiran [1987], p.25).

For the Pythagoreans however, numbers, meaning natural numbers, were "the first things in the whole of nature". Thus in Aristotle we read:

... since, then, all other things seemed in their whole nature to be modelled on numbers, and numbers seemed to be the first things in the whole of nature, they supposed the elements of numbers to be the elements of all things, and the whole heaven to be a musical scale and a number. And all the properties of numbers and scales which they could show to agree with the attributes and parts and the whole arrangement of the heavens, they collected and fitted into their scheme; and if there was a gap anywhere, they readily made additions so as to make their whole theory coherent (Metaphysica, 985b - 986a).

On the other hand,

the Greeks never identified a non-rational square root with a number (Dedron & Itard [1974], vol.2, p.102).

What was the context in which numbers and magnitudes were considered and what do we know of Pythagorean mathematics?

If we wished to know exactly what Pythagoras did and taught, we should have to travel back in time. This is because the sources of our information are at most about two thousand years old, while the age of Pythagoras was the sixth century B.C.. Indeed, those sources are late transcriptions of earlier writings. According to the Oxford Classical Dictionary [b], Pythagoras "wrote probably nothing (although works were later fathered on him) and already in Aristotle's day his life was obscured by legend". For us the *Vita Pythagorica* of Iamblichus [1818] is a major source but Iamblichus lived from *c.* 250-325 A.D. ([b], p.538), so the reliability of the text (let alone its content) cannot be guaranteed. Even the little which is told of Pythagoras by Aristotle (384-322 B.C.) was written originally more than a hundred years after Pythagoras's death (490 B.C., de Vogel [1966], p.21, 24). Again Aristotle's work, like Iamblichus', was transcribed and the originals are not now available, though in this case we have a number of (late) manuscripts.

It is impossible to say how much distortion has been introduced by this transcription process. There is no doubt that much has been. However, as in the case of the stories of Christ, a picture of a certain charismatic character with exceptional gifts, including those of leadership, does come through. Moreover, although the three biographies of Pythagoras which have been preserved in some form, in Diogenes Laertius' *Vitae Philosophorum* [1925], Iamblichus [1818] and Porphyry [1815-16] "are still considered as belonging rather to the genre of hagiography than to history" (de Vogel [1966], p.5)[3], nevertheless a picture of considerable coherence does emerge. In order to give some indication of the sort of individual Pythagoras was, we shall shortly turn to Plato's writings.

Plato was a Pythagorean in many of his views (Aristotle, *Metaphysica*, 987a-988a) and although Aristotle says Plato's philosophy "had peculiarities that distinguished it from the Italians [i.e. Pythagoreans]", in particular "in making the One and the Numbers separate from things, and his introduction of the Forms", it is reasonable to assume that the basic tenor of Pythagoras' philosophy in the old-fashioned sense of attitude to, and view of, the world was strongly reflected in Plato's philosophy.

Pythagoras' and the Pythagoreans' main interest was number in all its forms.

> ... *the so-called Pythagoreans, who were the first to take up mathematics, not only advanced this study, but also having been brought up in it they thought its principles were the principles of all things (Aristotle, Metaphysica, 985b).*

3. Ideas of mathematics

In ancient Greek mathematics the drawing of distinctions around the concept of number was somewhat different from what it is today. It is easy to forget this, though happily not everyone does. Nowadays we use the word "number" in many different ways, in particular we use numbers for counting and for measuring. We also use the concept of number in a more vague way – a large number of people, several pounds of beans (*cf.* chapter I). For the ancient Greeks of the fifth or fourth century B.C. "number" was used in ways that are, to us, so different that it is difficult for us to comprehend that there is even what Wittgenstein calls a "family resemblance" (Wittgenstein [1958], no. 68).

In the Philosophical Investigations, Wittgenstein says:

And for instance the kinds of number form a family ... Why do we call something a "number"? Well, perhaps because it has a — direct — relationship with several things that have hitherto been called number; and this can be said to give it an indirect relationship to other things we call the same name. And we extend our concept of number as in spinning a thread we twist fibre on fibre. And the strength of the thread does not reside in the fact that some one fibre runs through its whole length, but in the overlapping of many fibres. But if someone wished to say: "There is something common to all these constructions — namely the disjunction of all their common properties" — I should reply: Now you are only playing with words —.

According to Aristotle, "the so-called Pythagoreans, who were the first to take up mathematics ... thought its principles were the principles of all things" (*Metaphysica*, 986a). Even in Plato the pre-eminence of number is still evident. This is well illustrated by the following extract from the *Republic* (524e – 525b). Here Socrates is describing the education of the proposed philosopher-rulers:

"... the soul in perplexity, is obliged to rouse her power of thought and to ask: 'What is absolute unity?' This is the way in which the study of the one has a power of drawing and converting the mind to the contemplation of true being.

"And surely, he said, this occurs notably in the visual perception of unity; for we see the same things at once as one and as infinite in multitude?

"Yes, I said; and this being true of one must be equally true of all number?

"Certainly.

"And all arithmetic and calculation have to do with number?

"Yes.

"And they appear to lead the mind towards truth?

"Yes, in a very remarkable manner.

"Then this is a discipline of the kind for which we are seeking ... ."

Although number was the basis of the Pythagorean philosophical system, it is not clear that mathematics was regarded as a coherent whole. From the works of Anatolius[4] we learn:

The Pythagoreans are said to have given the special name **mathematics** *only to geometry and arithmetic; previously each had been called by its separate name, and there was no name common to both (see Thomas [1939], vol. I, p.2).*[5]

Elsewhere mathematics is said to have comprised four or five divisions. Plato speaks of the five mathematical studies: arithmetic, geometry, solid geometry, astronomy and harmony (or harmonics) (*Republic*, 525 f.). On the other hand Porphyry (232/3 - *c.* 305 A.D., O.C.D. [b], p. 864) cites Archytas, who lived in the first half of the fourth century B.C., listing four studies.

Let us now cite the words of Archytas the Pythagorean, whose writings are said to be mainly authentic. In his book **On Mathematics** *right at the beginning of the argument he writes thus:*

'The mathematicians seem to me to have arrived at true knowledge, and it is not surprising that they rightly conceive the nature of each individual thing; for, having reached true knowledge about the nature of the universe as a whole, they were bound to see in its true light the nature of the parts as well. Thus they have handed down to us clear knowledge about the speed of the stars, and their risings and settings, and about geometry, arithmetic and sphaeric, and, not least, about music; for these studies appear to be sisters (see Thomas [1939], vol.I, p.4 and note).

Now we shall only treat arithmetic and geometry. What is important is the view that there was a rather more tenuous nature in the connexion between arithmetic and geometry than there is for us today.

4. Pythagoras' education

According to Iamblichus, Pythagoras' teacher was Thales of Miletus (Iamblichus [1818], ch. II, p.6).[6]

Thales ... was born about B.C. 640 at Miletus, and died at the same place about B.C. 542. The main facts of his life are given by Diogenes Laertius ([1925], vol.I, p.23-47), who cites Apollodorus as authority for the birth of Thales in the 35th Olympiad, and Socrates, for his death in the 58th (Gow [1884], p.138-139).

Thales was a Phoenician by remote descent who had some interest in politics for Herodotus quotes an excellent political proposal made by Thales, "that the Ionians should set up a common centre of government at Teos, as that place occupied a central position; the other cities would continue as going concerns, but subject to the central government" (Herodotus I, 169). On the mathematical side, Thales is chiefly memorable for the following assertions which are attributed to him – all in the realm of geometry.

(1) The circle is bisected by its diameter.
(2) The angles at the base of an isosceles triangle are equal.
(3) If two straight lines cut one another, the opposite angles are equal.
(4) The angle in a semicircle is a right angle.
(5) A triangle is determined if its base and base-angles be given (practically Euc. I, 26) (Gow [1884], p.138-139).

(1) is "merely stated as a fact [by Euclid] in I, Def. 17. Thales therefore probably observed rather than proved the property" (Heath [1925], vol.I, p.131). But it is attributed to Thales by Proclus ([1970], p.124). Gow writes:

The language of Proclus also[7] seems to hint that Thales proved the proposition (2), our old friend, the **Pons Asinorum** by taking **two** equal isosceles triangles and applying them to one another as in Euc. I. 4, another case of experiment. But the theorems [(4) and (5)] are obviously incapable of such treatment [i.e. by super-position], and must have been supported either by deduction or at least by very wide induction. The last of them (Euc. I, 26) is attributed to Thales by Eudemus (Proclus, [1970], p.65), apparently on the ground that Thales invented a mode of discovering the distance of a ship at sea, in which the proposition was used (Gow [1884], p.141).

However, on (4) Diogenes Laertius writes:

Pamphila says that, having learnt geometry from the Egyptians, he was the first to inscribe in a circle a right-angled triangle, whereupon he sacrificed an ox. Others say it was Pythagoras, among them being Apollodorus the calculator (Diogenes Laertius [1925], p.25-27).

According to Proclus, Thales

first went to Egypt and thence introduced this study (geometry) into Greece. He discovered many propositions himself, and instructed his successors in the principles underlying many others, his method of attack being in some cases more general (i.e. more theoretical or scientific), in others more empirical (αισθητικωτερον, more in the nature of simple inspection or observation) (Proclus [1970], p.52).

Pythagoras, on the other hand, went to Thales first, according to Iamblichus.

But after he had attained the eighteenth year of his age, about the period when the tyranny of Policrates first made its appearance, foreseeing that under such a government he might receive some impediment in his studies, which engrossed the whole of his attention, he departed privately by night with one Hermodamas ... to Pherecydes, to Anaximander the natural philosopher, and to Thales at Miletus. He likewise alternately associated with each of these philosophers, in such a manner, that they all loved him, admired his natural endowments, and made him a partaker of their doctrines (Iamblichus [1818], ch.II, p.5-6).

Iamblichus goes on to say:

Indeed, after Thales had gladly admitted him to his intimate confidence, he admired the great difference between him and other young men, whom Pythagoras left far behind in every accomplishment. And besides this, Thales increased the reputation Pythagoras had already acquired, by communicating to him such disciplines as he was able to impart: and, apologizing for his old age, and the imbecility of his body, he exhorted him

115

Real Numbers

*to sail into Egypt, and associate with the Memphian and Diospolitan[8]
priests.   For he confessed that his own reputation for wisdom, was derived
from the instructions of these priests; but that he was neither naturally,
nor by exercise, endowed with those excellent prerogatives, which were so
visibly displayed in the person of Pythagoras.   Thales, therefore, gladly
announced to him, from all these circumstances, that he would become the
wisest and most divine of all men, if he associated with these Egyptian
priests (ibid.).*

Pythagoras

*spent therefore two and twenty years in Egypt, in the adyta of temples,
astronomizing and geometrizing, and was initiated, not in a superficial or
casual manner, in all the mysteries of the Gods, till at length being taken
captive by the soldiers of Cambyses, he was brought to Babylon.   Here he
gladly associated with the Magi, was instructed by them in their venerable
knowledge, and learnt from them the most perfect worship of the Gods.
Through their assistance likewise, he arrived at the summit of arithmetic,
music, and other disciplines; ... (ibid., ch.IV, p.9).*

Whether this account is accurate or not, it is known that precise astronomical observations were made in Babylon[9] and Thales is believed to have predicted the year of a solar eclipse (see Herodotus I.73).   Moreover, Thales is designated as being "of Miletus" (see above) and Miletus is in Asia Minor.   He therefore came from a place as close to Babylon as Italy, Pythagoras's main abode (de Vogel [1966], p.24), was to his birthplace, Samos (a small island not far from Miletus) (Philip [1966], p.185).   *Prima facie*, therefore, it is possible that Thales was familiar with some Babylonian mathematics, but we shall see below that there are difficulties in arguing that either Thales or Pythagoras was *very* familiar with Babylonian mathematics.

According to Iamblichus ([1818], ch. II, p.6), Pythagoras had brought his knowledge of some parts of arithmetic and geometry from Egypt, but whether Pythagoras actually visited Egypt or Babylon is questionable.   Thus J.A. Philip writes:

*The voyages of Pythagoras were a favourite theme of the biographical
tradition.   In early times philosophers and sophists, as they shared the
name of "sophist", shared also the characteristic of travelling from city to
city and from country to country.   In that he remained all his life in
Athens, Socrates was an exception to the rule.   His pupil Plato, however,
in his first voyage to Sicily undertook what was a voyage of instruction in
the strict sense, and thereafter it was common for philosophers to embark on
a journey abroad to learn what there was to be learned in foreign lands.
They naturally imagined their predecessors to have made similar journeys.*

But there are many more questions to be asked as far as both Egypt and Babylon are concerned.   First let us turn to Egypt.

5. Egyptian mathematics

It is ironic that although Western mathematics has remained incredibly dependent on Greek mathematics, indeed on Euclid's Elements, no original documents of very early Greek mathematics have survived, while in the case of Egypt in the same period we have less mathematics and more documents[10] Even here these only number a handful.

The documents we do possess indicate an overwhelming concern with mensuration and, indeed, very "concrete" mathematics. van der Waerden judges Egyptian geometry as "not a science in the Greek sense of the word, but merely *applied arithmetic*" (van dcr Waerden [1975], vol.I, p.31, 36). Peet, in his edition of the Rhind Mathematical Papyrus, perhaps our most important surviving Egyptian mathematical work, puts it better than some later writers:

The outstanding feature of Egyptian mathematics is its intensely practical character. This is not peculiar to mathematics, for it is typical of all the sciences in Egypt. As Plato alone of the Greeks seems to have realized[11] the Egyptians were essentially a "nation of shopkeepers", and interest in or speculation concerning a subject for its own sake was totally foreign to their minds.

To realize this we have only to take a glance through the problems of the Rhind Papyrus. Here everything is expressed in concrete terms. The Egyptian does not speak or think of 8 as an abstract number, he thinks of 8 leaves or 8 sheep. He does not work out the slope of the sides of a pyramid because it interests him to know it, but because he needs a practical working rule to give to the mason who is to dress the stones ... If he resolves $\frac{2}{13}$ into $\frac{1}{8} + \frac{1}{52} + \frac{1}{104}$ it is not because this fact in itself appeals in any way to his curiosity, but simply because sooner or later he will come across the fraction $\frac{2}{13}$ in a sum, and since he has no machinery for dealing with fractions whose numerators are greater than unity he will then urgently need the resolution above stated.

Perhaps it is in keeping with this attitude that there is in our papyrus [Rhind] practically no instance of the use of a general formula, each case being worked out on its own merits, and cases which to us seem analogous being sometimes dealt with by totally different methods (Peet [1923], p.10).

The actual nature of the problems given in the Rhind Mathematical Papyrus makes one question just how "practical" the Egyptian mathematicians were. It is very noticeable that all the answers work out neatly, though even today most mathematics textbooks of the present era have this same tendency. However, there appears to be a very modern, that is twentieth century, tradition that Egyptian mathematics was not "only practical" even in this restricted sense. Gillings dissects this view nicely. Amongst other justified comments of his we find:

... a sober-minded person as H.W. Turnbull writes, "Their land
surveyors were known as rope stretchers, because they used ropes with knots
or marks at equal intervals to measure their plots of land. By this simple
means, they were able to construct right angles, for they knew that three
ropes of lengths three, four, and five units respectively, could be formed
into a right-angled triangle.[12]

It is, however, nowhere attested that the ancient Egyptians knew even
the very simplest case of Pythagoras' theorem! But Turnbull goes further:
"As Professor D'Arcy Thompson has suggested, the very shape of the Great
Pyramid indicates a considerable familiarity with that [sic] of the regular
pentagon. A certain obscure passage in Herodotus, can, by the slightest
literal emendation, be made to yield excellent sense. It would imply that
the area of each triangular face of the Pyramid, is equal to the square of
the vertical height. If this is so, the ratios of height, slope, and base,
can be expressed in terms of the golden section, or the ratio of a circle
to the side of the inscribed decagon."

I am unable to understand exactly what Turnbull means by this last
sentence. But whatever it means, with further slight emendations, the
dimensions of the Eiffel Tower or Boulder Dam could be made to produce
equally vague and pretentious expressions of a mathematical connotation
(Gillings [1972], p.238).

Egyptian mathematics with its supposed connexions with the pyramids has
attracted wild speculation, while the actual results contained in the Rhind
Mathematical Papyrus are very interesting and substantial (see below).

Thus there is, to our knowledge, nothing even remotely approaching
Pythagoras' theorem. Gillings has five quotations ([1972], App. 5, p.242).
All indicate the falseness of another modern tradition, namely that the
Egyptians at least knew of the right-angled triangle with sides 3, 4 and 5,
but one author he quotes only indirectly is Peet, who says:

... an interesting problem is raised by Democritus' reference to the
harpedonaptai of Egypt. The philosopher boasted that no one of his time
had surpassed him in constructing figures from lines and in proving their
properties, not even the so-called harpedonaptai of Egypt. Who were these
harpedonaptai? More than one historian of mathematics has supposed that
they were land-measurers. The literal meaning of the word is
"rope-stretchers", and it is suggested [by Heath [1925], vol.I, p.122] that
they were acquainted with the fact that a triangle whose sides were 3, 4 and
5 contained a right-angle, and that they constructed right-angles
accordingly, as did the Chinese and the Indians.

For this last statement I can find no foundation whatsoever: nothing in
Egyptian mathematics suggests that the Egyptians were acquainted even with
special cases of Pythagoras' theorem concerning the squares on the sides of
a right-angled triangle. That the harpedontaptai were land-measurers on
the other hand is most probable, indeed we can even see such persons at work
in the pictures on the walls of Egyptian tombs (Peet [1923], p.31-32).

Now the Egyptian Rhind Papyrus (BM 10057 and 10058) is dated about 1650
B.C.,[13] so it is possible that geometry did develop significantly in Egypt in

the millenium preceding the time of Pythagoras. Moreover, Heath gives details of other documentation for the development of geometry in Egypt, but says that "statements [Heath quotes] may all be founded on the passage of Herodotus, and Herodotus may have stated as his own inference what he was told in Egypt" (Euclid [1925], vol.I, p.121). In Herodotus we read:

It was this king [Sesostris]¹⁴, moreover, who divided the land into lots and gave everyone a piece of equal size, from the produce of which he exacted an annual tax. Any man whose holding was damaged by the encroachment of the river would go and declare his loss before the king, who would send inspectors to measure the extent of the loss, in order that he might pay in future a fair proportion of the tax at which his property had been assessed. Perhaps this was the way in which geometry was invented, and passed afterwards into Greece — for knowledge of the sundial and the gnomon and the twelve divisions of the day came into Greece from Babylon (Herodotus, II, 109).

Thus, in the millenium preceding Pythagoras, Egyptian mathematics as we know it is concerned with specific contexts, as is all mathematics in this period. Later we shall see that it was concerned with calculations in what to us is a geometrical context but there is no record of anything approaching even the geometric assertions attributed to Thales (see above, §.4).

6. Babylonian mathematics

Turning now to Babylon, the situation is somewhat different. The "oldest preserved document in number theory ... tabulates the answers to a problem containing Pythagorean numbers (or Pythagorean triangles)" (Neugebauer & Sachs [1945], p.37). This document is the cuneiform tablet Plimpton 322 of the Plimpton collection in Columbia University, New York. "It falls in the period between 1900 and 1600 B.C." (ibid., p.39) and is thus from the same era as the Rhind Mathematical Papyrus. Now, according to Neugebauer and Sachs, the "terminology [of cuneiform tablets in general] is geometrical, [but] the whole treatment is strongly algebraic" (ibid., p.37). 3, 4, 5-triangles specifically occur: thus Plimpton 322 line 13 contains the numbers 45, 75 and Neugebauer and Sachs interpolate the 60, while in the British Museum tablet BM 34568 we have (from Neugebauer's translation):

[4 is the leng]th and 5 the diagonal. What is the breadth? Its size is not known. 4 times 4 (is) [1]6. 5 times 5 (is) 25. You take 16 (from) 25 and there remains 9. What times what shall I take (in order to get) 9? 3 times 3 (is) 9. 3 (is) the breadth (Neugebauer [1935], vol.III, p.17).

There are other similar examples given by Neugebauer. Thus numerical versions of Pythagoras' theorem were certainly known from the Old Babylonian period up to 1600 B.C. up to the Seleucid period which started about 300 B.C. (Neugebauer [1957], p.14). However, there is no evidence at all of what we would call well-developed general methods, though specific problems

119

may be construed as general. Thus on BM 34568 we find Neugebauer commenting:

> *Examples 14 and 17 are worked out according to (4), as usual, in concrete numbers. But of greater fundamental interest is the fact that the formula (4) in No. 18 is written completely without particular numbers and therefore is really a **general formula** (Neugebauer [1935], vol.III, p.21).*

But all of these are in a tradition which is concerned with mensuration. There appears to be no evidence at all of synthetic geometry in Babylon or Egypt before the time of Thales (*cf.* e.g. Neugebauer [1957], p.44).

Thus synthetic geometry seems to have sprung up very rapidly indeed according to the evidence we have. It is claimed by van der Waerden:

> *... what is characteristic and absolutely new in Greek mathematics, is the advance by means of demonstration from theorem to theorem. Evidently, Greek geometry has had this character from the beginning, and it is Thales to whom it is due (van der Waerden [1975], p.89).[15]*

However, while the claim that Greek mathematics right from the time of Thales proceeded from theorem to theorem by demonstration is not as clear to the present author as it appears to be to van der Waerden, there does appear to be a clear qualitative difference between early Greek geometry and the early geometry of Egypt and Babylon. This is, at least apparently, very similar to the difference we noted above between Greek and Chinese mathematics: Greek mathematics is conducted in terms of proof; Chinese mathematics is concerned with computing answers. In Egypt, Babylon and countries further East, geometry is always intuitively connected with mensuration and calculation in all the records we possess. In Greece, from the time of Euclid at least, geometry seems to have existed in a deductive form based on theorems and proofs much closer to that part of pure mathematics as we know it today.

7. Approximations

So far, in all the mathematics we have discussed, we have avoided mention of irrationals or even fractions. In the time of Pythagoras and even until later

> *... no example of simple division nor any rules for division are found in Greek arithmetical literature. The operation must have been performed by subtracting the divisor or some easily ascertained multiple of the divisor from the dividend and repeating this process with the successive remainders. The several quotients were then added together. But the Greeks had no name for a **quotient** and did not conceive the result of a division as we do. To a Greek 5 was not the **quotient** of $\frac{35}{7}$. The*

operation did not discover the fact that 5 times 7 is 35 but that a seventh part of 35 contains 5, and so generally in Greek a division sum is not stated in the form "Divide **a** *by* **b***", but in the form "Find the* **b***th part of* **a** *(Gow [1884], p.51).*

The whole thrust of Pythagorean number theory as we know it (principally through Euclid) is in terms of submultiples. Thus in the Definitions at the beginning of Euclid Book VII we read (in translation):

3. *A number is a* **part** *of a number, the less of the greater, when it measures the greater;*
4. *but* **parts** *when it does not measure it.*
5. *The greater number is a* **multiple** *of the less when it is measured by the less (Euclid [1925], vol.II, p.279).*

Heath is generally followed in believing these definitions to be Pythagorean (see Thomas [1939], vol.I, p.66 note). Moreover, Nicomachus (second century A.D., see Nicomachus [1926], p.71) also, in discussing the concept of number as a multitude (Diophantos [1893/85], p.190), puts his whole discussion in the context of a "Pythagorean doctrine". Iamblichus, who wrote a commentary (Iamblichus [1894]) on Nicomachus, referred to the writings of many "ancient Pythagoreans ... and all the books which they published" as still being extant (Iamblichus [1818], ch. XXIII, p.55-56). Thus we have grounds for regarding Heath as expounding Pythagorean doctrine when he says:

By a part Euclid means a submultiple, as he does in V. Def. I, with which definition this one is identical except for the substitution of number (ἀριθμός) for **magnitude** *(μέγεθος); cf. note on V. Def. I. Nicomachus uses the word "submultiple" (ὑποπολλαπλάσιος) also. He defines it in a way corresponding to his definition of multiple [see note on Def. 5 below] as follows (I. 18, 2): "The submultiple, which is by nature first in the division of inequality (called) less, is the number which, when compared with a greater, can measure it more times than once so as to fill it exactly (πληροῦντως)." Similarly sub-double (ὑποδιπλάσιος) is found in Nicomachus meaning* **half***, and so on (Euclid [1925], vol.II, p.280)*

and

The definition of a **multiple** *is identical with that in V. Def. 2, except that the masculine of the adjectives is used agreeing with ἀριθμός understood instead of the neuter agreeing with μέγεθος understood. Nicomachus (I,.18,I) defines a multiple as being "a species of the greater which is naturally first in order and origin, being the number which, when considered in comparison with another, contains it in itself completely more than once" (ibid.).*

Thus the whole basis of the Pythagorean treatment of what we could call "fractions" appears to consist in a reduction to integral multiples of a basic unit. In Egypt, however, a tremendous procedure is needed to deal with (what to us is) the simplest addition of fractions; about 20% of the Rhind Mathematical Papyrus is occupied with tables for 2/n (n an odd number from 5 to 101, inclusive) thus indicating that adding, say, 1/7 to

1/7 was non-trivial to the Egyptians. In this example 1/7 + 1/7 = 1/4 + 1/28, as the Egyptians do not use numerators greater than one (with the single exception of 2/3) and always insist on sums of distinct submultiples (Peet [1923], p.15). Whether it was so difficult for the Babylonians is difficult to say, for non-sexagesimal fractions do not seem to be very common in cuneiform texts.

In Egypt, as we noted above, mathematics was of a very practical bent and although we have much less evidence than we have for Babylonian mathematics, the difficulties the Egyptians experienced with the representation of fractions argues against their concern with any irrational number, though we can say that $4(8/9)^2$ was taken as a value of π in the Rhind papyrus in finding the area of a circle (Rhind Mathematical Papyrus, problems S.41-43, 48, 50; (Peet [1923], p.81, 89)). Further, as we have argued above, the idea of anything related to Pythagoras' theorem in Egypt seems to be a late fabrication. However, we do not know of any actual approximate calculations in Egypt where exact answers would require the use of irrationals.

In Babylon the situation is quite different. The cuneiform tablets from Babylon that have been studied "can be classified into two major groups: 'table texts' and 'problem texts' (Neugebauer [1957], p.30). Some of these latter are concerned with calculating the lengths of sides of geometric figures (e.g. BM 34568 above, Neugebauer [1935], p.17-19), but nowhere is there anything seen to be close to what we regard as the geometry of Thales and his successors. On the other hand, there are two very important features of the tablets. One is the presence of tables of reciprocals and the other the calculation of approximations to irrational numbers.

The Babylonians of about 1600 B.C. used a sexagesimal system (i.e. used 60 as a base instead of the 10 of the decimal system cf. Ch. III, §1). However, they did not have a well-defined place system, thus a number (now represented by) 1,30 could be interpreted as 90 or 1 1/2 or 1/40, etc. Presumably the context (not necessarily written down) determined the desired value. However, because $60 = 2^2.3.5$ has a large number of distinct factors (12, in fact), the reciprocals of many numbers with only multiples of 2, 3 and 5 as factors can be calculated exactly. Neugebauer and Sachs ([1945], p.11 f.) present many tables of reciprocals from cuneiform texts. These extend the numbers considered beyond simple ones like 1/6 = 10, 1/12 = 5 to ones even more complicated than 1/8,38,24 = 6,56,40 (1/0.144 = 6.94 in modern notation). (From the reverse of CBS 29.13.21 col. III quoted in Neugebauer and Sachs [1945], p.14.) Although these examples give neat fractions, YBC 10529 (*ibid.*, p.16) contains a list in which the reciprocals are approximations to three or four sexagesimal places. Thus approximations of reciprocals were used in Babylon.

The second important feature is the approximation of irrational numbers. Perhaps the most striking is one example for $\sqrt{2}$, though YBC 7302 (*ibid.*, p.44) gives the approximation 3 for π. The tablet YBC 7289 (*ibid.*, p.42) shows a square of side 30 (= 1/2) and diagonal 42,25,35 = 30 × (1,24,51,10). This last factor (1,24,51,10) is a very good estimate of $\sqrt{2}$ = 1.4142 ... .

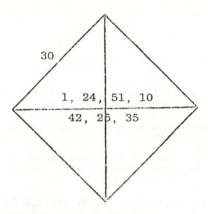

30

1, 24, 51, 10

42, 25, 35

(Neugebauer and Sachs mention in a footnote that there is a value of 1,25 [= 1.416] in a later, Seleucid, tablet (*ibid.*, p.43).) It is impossible, without further evidence, to show how such a good value was obtained, but Neugebauer and Sachs give the following proposal which has (at the very least) the merit of giving exactly 1, 24, 51, 10.

The procedure in question consists in the alternating approximation of a by arithmetic and harmonic means of previously found approximations. Let α_1 be any approximation of $\sqrt{a}$ such that $\alpha_1 > \sqrt{a}$. Then $\beta_1 = \dfrac{a}{\alpha_1}$ is also an approximation of $\sqrt{a}$ but deviates from the true value in the opposite direction because it follows from $\alpha_1 > \sqrt{a}$ that $\beta_1 = \dfrac{a}{\alpha_1} < \sqrt{a}$. We now derive a new pair of approximations, α_2 and β_2, by

$$\alpha_2 = \frac{\alpha_1 + \beta_1}{2} \qquad \beta_2 = \frac{a}{\alpha_2}$$

and continue this process by computing

$$\alpha_3 = \frac{\alpha_2 + \beta_2}{2} \qquad \beta_3 = \frac{a}{\alpha_3}$$

etc.[16] *It is evident that all the α's are greater than $\sqrt{a}$, all β's less than $\sqrt{a}$, but that each step diminishes the difference between corresponding approximations. We apply this method to $\sqrt{2}$ by starting with $\alpha_1 = \dfrac{3}{2}$ as the first rough approximation (α_1^2 would be 2;15). Then we obtain for the first pair*

$$\alpha_1 = 1;30 \qquad \beta_1 = \frac{2}{1;30} = 1;20.$$

Real Numbers

The next step leads to

$$\alpha_2 = \frac{1}{2}(1;30 + 1;20) = 1;25$$

$$\beta_2 = \frac{2}{1;25} = 1;24, 42, 21,\dots .$$

Here we have already reached the above mentioned value 1.25 as an approximation to $\sqrt{2}$. The very next step leads to

$$\alpha_3 = \frac{1}{2}(1;25 + 1;24,42,21,\dots) = 1;24,51,10,\dots$$

the value given in (1).[17] The fact that both values for $\sqrt{2}$ found in our texts are links of the same chain seems to be a rather strong argument in support of our explanation[18] (ibid., p.39).

Thus the Babylonians were either capable of calculating lengths without having to reduce figures to multiples of a basic unit or else did not strive to get complete accuracy in their calculations. In either case the arithmetic involved in Babylonian mathematics is markedly different from the Greek arithmetic of natural numbers before the discovery of the irrationality of $\sqrt{2}$. This appears to support the view that Pythagoras did not learn mathematics from Babylon or at least was not strongly influenced by Babylonian mathematics, especially as these tablets date from both the Old Babylonian period (*circa* 1900 B.C. to 1600 B.C.) and the later period (700 B.C. - A.D. 0.). We have no evidence at all that Pythagoras was involved in the Babylonian style of mathematics.

Thus we see that both in Babylon and in Egypt there was concern with approximate values, but that this was not so in Greece. For our present concerns it is immaterial whether the Babylonians and Egyptians were aware of the possibility of real numbers which could not be finitely expressed as fractions (or finite decimals or sexagesimals) and which therefore could never be written down exactly.

On the other hand the

Greeks followed the ancient plan of avoiding, by the use of submultiples, the difficulty of computing with fractions; but in due time [about the 3rd century B.C.] the need for a fraction symbolism became so apparent that they developed a system that served their purposes fairly well (Smith [1923], vol.II, p.214).

The natural numbers seem to have served the purposes of the world until about the beginning of the historic period. Men broke articles and spoke of the broken parts, but even after weights came into use it was not the custom to speak of such a fraction as 3/4 of a pound. The world avoided difficulties of this kind by creating such smaller units as the ounce and then speaking of the particular number of ounces (ibid., p.208).

Even in the fifth century, the Greeks were using a system of literals for numerals which did not employ a place notation and indeed it is not clear how much they were using letters for numbers at the time of

Pythagoras.   Thus   Heath   ([1921],   vol.I,   p.33   f.)   gives   various   not
completely   conclusive   arguments   for   dates   from   750   B.C.   on   for   the
introduction of such symbols and notes that

> *It was a long time before the alphabetic numerals found general*
> *acceptance.   They were not officially used until the time of the Ptolemies*
> *[3rd century B.C.] (ibid., p.34).*

According   to   Heath   is   it   not   until   Aristarchus   (*c.*   310-230   B.C.)   that
we find fractions used in Greek mathematics (*ibid.*, p.43).

8. Figured numbers

What   we   believe   to   be   Pythagorean   arithmetic,   as   handed   down   to   us   by,
for   example,   Nicomachus   ([1926],   ch.XII,   p.246   f.),   certainly   depends
heavily   on   the   representation   of   numbers   by   arrays   of   points   (or   pebbles)
(*cf.* above, chapter II, §2).

> *For example, the writing of one unit by means of one alpha will be the*
> *sign for 1; two units side by side, that is, a series of two alphas, will*
> *be the sign for 2; when three are put in a line it will be the character*
> *for 3, four in a line for 4, five for 5, and so on.   For by means of*
> *such a notation and indication alone could the schematic arrangement of the*
> *plane and solid numbers mentioned be made clear and evident, thus:*

> *The number 1,*   α
> *The number 2,*   α α
> *The number 3,*   α α α
> *The number 4,*   α α α α
> *The number 5,*   α α α α α

and further in similar fashion (ibid., ch. VI, p.237).

> *To illustrate and classify, linear numbers are all those which begin*
> *with 2 and advance by the addition of 1 in one and the same dimension; and*
> *plane numbers are those that begin with 3 as their most elementary root and*
> *proceed through the next succeeding numbers.   They receive their names also*
> *in the same order; for there are first the triangles, then the squares, the*
> *pentagons after these, then the hexagons, the heptagons, and so on*
> *indefinitely,... (ibid., ch. VII, p.239).*

For a further example we have:

> *The pentagonal number is one which likewise upon its resolution into*
> *units and depiction as a plane figure assumes the form of an equilateral*
> *pentagon.   1, 15, 12, 22, 35, 51, 70, and analogous numbers are examples.*
> *Each side of the first actual pentagon, 5, is 2, for 1 is the side of the*
> *pentagon potentially first, 1; 3 is the side of 12, the second of those*
> *listed; 4, that of the next, 22; 5, that of the next in order, 35, and 6 of*
> *the succeeding one, 51, and so on.   In general the side contains as many*
> *units as are the numbers that have been added together to produce the*

pentagon, chosen out of the natural arithmetical series set forth in a row (ibid.,ch.X, p.234).

For example, 5 and 12 are represented as

The theory of proportions is regarded even by as late an exponent as Nicomachus as being something quite distinct.   At the beginning of chapter XXI he says:

After this it would be the proper time to incorporate the nature of proportions, a thing most essential for speculation about the nature of the universe and for the propositions of music, astronomy, and geometry, and not least for the study of the words of the ancients, and thus to bring the **Introduction to Arithmetic** *to the end that is at once suitable and fitting (ibid., p.264).*

Further, all the proportions he discusses are presented in terms of integers. Indeed, the whole book contains no reference to irrationals.   Of course it could be argued that since the book is entitled *Introduction to Arithmetic* there is no need to mention numbers other than integers.

Thus we can draw a clear distinction here between the mathematics of Nicomachus (meaning, in our terminology, arithmetic and figured numbers) and the theory of mensuration and proportion (as treated in, say, Euclid's Books VII, VIII and IX).   If we regard Nicomachus as proceeding in the Pythagorean spirit as he appears to do (*cf. ibid.*, p.254, 266, 283), then it is reasonable to conclude that the orientation of the Pythagoreans was very much in terms of representation of numbers by geometrical forms.   Such a geometrical representation is, in modern mathematical terms, closely akin to working in a discrete space rather than a continuous one.   Now we know that it is quite possible to do topology in such a space[19] and it follows that there is no need, from the philosophical point of view, to insist on a continuous space.   It therefore also follows that the move from discrete geometry (if we may label Nicomachus' graphical presentation of number in that way) to the comparison of lengths is just as inessential.   Indeed, as we have seen, Euclid and the Greeks generally dealt separately with numbers and magnitudes.

9. Magnitudes and ratios

Magnitudes are treated in Euclid's Book V and then numbers separately in Book VII, while the more complicated treatment of incommensurables is reserved for Book X.   Heath (Euclid [1925], vol.II, p.113) regards it as "remarkable" that Euclid treats numbers and magnitudes separately.   In the

Theaetetus. *Theodorus was writing out for us something about roots, such as the sides of squares three or five feet in area, showing that they are incommensurable by the unit: he took the other examples up to seventeen, but there for some reason he stopped. Now as there are innumerable such roots, the notion occurred to us of attempting to find some common description which can be applied to them all.*

Socrates. *And did you find any such thing?*

Theaetetus. *I think that we did; but I should like to have your opinion.*

Socrates. *Let me hear.*

Theaetetus. *We divided all numbers into two classes, those which are made up of equal factors multiplying into one another, which we compared to square figures and called square or equilateral numbers; — that was one class.*

Socrates. *Very good.*

Theaetetus. *The intermediate numbers, such as three and five, and every other number which is made up of unequal factors, either of a greater multiplied by a less, or of a less multiplied by a greater, and, when regarded as a figure, is contained in unequal sides; — all these we compared to oblong figures, and called them oblong numbers.*

Socrates. *Capital; and what followed?*

Theaetetus. *The lines, or sides, which have for their squares the equilateral plane numbers, were called by us lengths; and the lines whose squares are equal to the oblong numbers, were called powers or roots; the reason of this latter name being, that they are commensurable with the former not in linear measurement, but in the area of their squares. And a similar distinction was made among solids* (Theaetetus, 147d-148b).

The *Theaetetus* is thought to have been written about 368 B.C. (see edition of Jowett [1953], vol. III, p. 192), in other words, about 200 years after Pythagoras' birth. Given that Theodorus starts off with $\sqrt{3}$ and $\sqrt{5}$ and does not mention $\sqrt{2}$, it is reasonable to assume that the incommensurability of $\sqrt{2}$ was well known at that time. Conflicting views on the name of the exact discoverer persist and even the approximate date of the original discovery continues to be in doubt. However, ingenious arguments for the discoverer being Hippasus and the date being about 450 B.C., and certainly in the latter half of the fifth century B.C., have been made. The most detailed account is given in Knorr [1975] (see p.37-40 for the dating). Knorr's book is most useful for its references and the reader is recommended to consult as many of these as possible. I found the book impenetrable and inconsistent. Thus, after dismissing the evidence of Iamblichus and others (p.21 f.), he ultimately concludes that the accounts of neo-Pythagoreans including Iamblichus give *on the whole a plausible account of the political and intellectual climate surrounding that discovery [of incommensurables] (Knorr [1975], p.49).*

It seems reasonably clear that the discovery took place first in the context of measuring the diagonal of a square (see Heath [1921], *I*, p.90-91). This is the proof perhaps referred to by Aristotle:

... the fallacy is obvious; as for example that if the diagonal of a square is taken to be commensurable, odd numbers are equal to even ones (Prior Analytics i. 23, 41a).

The more modern version proceeds: Suppose $\sqrt{2} = m/n$, where m, n are positive integers with no common factor. Then $m = n\sqrt{2}$ and $m^2 = 2n^2$. Hence 2 divides m^2 and therefore 2 divides m. But then $m = 2p$, say, and therefore $(2p)^2 = 4p^2 = 2n^2$ and $2p^2 = n^2$. The same argument now yields 2 divides n. Therefore 2 divides both m and n contrary to our hypothesis. Thus $\sqrt{2}$ is not expressible as m/n, that is to say, $\sqrt{2}$ is irrational.[20]

This proof can be found in almost this form in Euclid book X, Proposition 117. However, it is generally agreed that this is a later addition to Euclid.

It is possible that this discovery came from consideration of dot figures or figured numbers (see above, chapter II, §2). Thus von Fritz ([1945], p.254) points out how figured numbers corresponding to integers can be used to demonstrate a significant difference between "oblong" and "square" numbers, that is, numbers representable only by configurations of unequal sides, as for example ° ° ° / ° ° ° and those which can be represented by squares, for example ° ° / ° ° . As we have seen, this is very much in the Pythagorean tradition (see §8 above). It is by no means impossible that by taking larger and larger squares, people came to realize that the super-position of a row of similarly spaced dots along the diagonal never exactly fits the diagonal. Thus one may, for a start, consider the following figures (where we have marked the diagonal dots with crosses and the distance between crosses is the same as between horizontally or vertically adjacent dots):

von Fritz, however, goes on to carry this style into geometric figures.

Almost immediately after the definitions of number cited above from Euclid book VII, we find the Euclidean algorithm.

Proposition I.   Two unequal numbers being set out, and the less being continually subtracted in turn from the greater, if the number which is left never measures the one before it until an unit is left, the original numbers will be prime to one another.

Proposition II.   Given two numbers not prime to one another, to find their greatest common measure (Euclid [1925], vol.II, p.296, 8).

The essence of the proofs is perhaps aptly described as by means of measuring off lengths representing numbers against each other.   The use of this technique is what von Fritz suggests, but he then gets involved in an infinite process.   This introduction of an infinite process seems both unnecessary and inappropriate.

Certainly Aristotle (see chapter II, §1above) was familiar with notions of infinite processes and Zeno's paradoxes are presented by Aristotle (*Physics* 9, 234b-261b).   He, however, clearly points out that there is no cause to use Zeno's arguments for proving there are incommensurables.   In the *Prior Analytics* (65b, 16-21) we read:

For to put that which is not the case as the case, is just like this: e.g. if a man, wishing to prove that the diagonal of the square is incommensurate with the side, should try to prove Zeno's theorem that motion is impossible, and so establish a **reductio ad impossibile***: for Zeno's false theorem has no connexion at all with the original assumption (Aristotle,* **Prior Analytics** *65b).*

Thus Aristotle is clear in separating infinite processes and arguments from the existence of irrationals.   It is therefore surprising to find that in von Fritz's arguments involving the ratio of the diameter to the side of a regular pentagon, von Fritz carries this on into an infinite process:

... if one looks at the pentagram or at a regular pentagon with all its diameters filled in [see figure on the next page] — and we have seen that the Pythagoreans were interested in diameters — the fact that the process of mutual subtraction goes on infinitely, that therefore there is no greatest common measure, and that hence the ratio between diameter and side cannot be expressed in integers however great, is apparent almost at first sight.   For one sees at once that the diameters of the pentagon form a new regular pentagon in the centre, that the diameters of this smaller pentagon will again form a regular pentagon, and so on in an infinite process (von Fritz [1945], p.257).

However, we could also observe that we get similar figures.   So in a sense we are getting back to the same situation: we are certainly getting back to the same ratio.   This is what happened in the odd-equals-even proof of Euclid X, 117.

Despite all this there never was and is even now no necessity to introduce infinity in any way in introducing incommensurable numbers or lengths.   The Greeks treated incommensurable lengths without doing so and

131

Aristotle (see above) actually pointed out that it was not necessary to introduce infinite considerations. Finally it is interesting to note how we use finite descriptions for specific interesting numbers which do not have finite presentations as decimals, for example $\sqrt{2}$, $\sqrt{3}$, π, e. and so on. Even though there are connotations of infinity in working out the decimal values of such numbers they can all be defined without bringing in the concept of infinity (Cf. also Wilder, [1973], p.124).

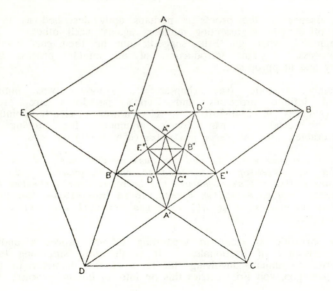

The question of why infinite processes and considerations enter into the subject of irrationals is what we shall take up in the next chapter.

Chapter VI

The Totality of Real Numbers

1. Introduction

We think nowadays of real numbers in general, and irrational numbers in particular, as *infinite* decimals.[1]   As we pointed out in the last chapter, there is no necessary connexion between infinite processes and incommensurable lengths.   However, if one is trying to use the Euclidean algorithm then the process of applying this algorithm can go on forever – as in the case of trying to find the common measure of 1 and $\sqrt{2}$.   Further, given an arbitrary length one can approximate it by taking units which, at each stage, are (say) one tenth of the size of the units at the previous stage.   This is how we construe decimal measures.   In fact this needs another axiom, known as Archimedes' axiom (see below).   Once we have these decimal measures, it is clear that we immediately have an infinite collection.

This idea, which gives rise to an infinite collection, was known to the Greeks and we shall shortly look at the comments of one of the Greeks: Proclus.   The idea was also well known and used in the East, in particular in India and China.   Development of the idea led to the question of comprehending *all* incommensurable lengths and this ultimately gave rise to the concept of all real numbers.   The difference between taking arbitrary lengths (or any other sort of magnitudes) and assigning decimal numbers to them was something which only started to develop in the work of Descartes and then was taken further in the work of nineteenth century mathematicians.

In the West it is not until we come to Stevin that we find the idea of a decimal number with an arbitrary (but still finite) number of places, but in the East this idea was current by, at the latest, 100 A.D..   The reason for this is that the algorithmic and practical approach in the East did not bring in the same philosophical problems as the Western approach through proof.[2]

The problems here include: what is the status of the "line"? and what is the correspondence between lengths and numbers?   We shall not say much about the former, but shall dwell on the latter.

In order to put the first problem into sharp relief we give two quotations.

A carpenter, using a metal ruler, draws a pencil line across a board to use as a guide in cutting   The line he has drawn is a physical thing; it is a deposit of graphite on the surface of a physical board (Davis and Hersh [1981], p.126).

133

Every time one draws the diagonal of a square, one has completed an infinite process of a type subtler than that pointed out by Zeno (Knorr [1975], p.39).[3]

The first accords with our experience and one thinks naturally of measuring, with a certain degree of precision, the length of such a line. The second shows confusion between the abstract definition and nature of a line and the physical process of drawing it.

Drawing or thinking of a line - it makes no difference which - one does consider the possibility of determining its length by a number. Considering these numbers one expects to be able to draw or think of a line whose length is *any* such number.   In this way we establish a correspondence between lengths and numbers which allows us to arithmetize, that is, to reduce the study of lengths (or any other kind of magnitude) to that of numbers.   This process was finally clarified in the nineteenth century by, first, Ch. Méray and then many others.   The explicit axiom required   here is due to Cantor and is sometimes known as Cantor's axiom.

Finally, although the work in the nineteenth century gave a sound axiomatic basis for identifying magnitudes and numbers, and removed the difficulties that had been caused by the importation of infinitesimals to do calculus,   work in the middle of the twentieth century led by Abraham Robinson[4] has in fact rehabilitated infinitesimals. ... These had been banished from mathematics for nearly a century but Robinson produced the notion of different, so-called "non-standard models" of the real numbers. Given the progress of history, it is possible this may lead to a different concept, possibly several different concepts, of real number in the future.

We now take up these authors and others in more detail.

2. The Greeks

Consider the following three quotations from Proclus' *Commentary on the First Book of Euclid's Elements*.

To find the principles of mathematical being as a whole, we must ascend to those all-pervading principles that generate everything from themselves; namely the Limit and the Unlimited (Proclus [1970], p.4, footnote of Morrow omitted).

... This is why in these orders of being these are ratios proceeding to infinity, but controlled by the principle of the Limit (ibid., p.5).

... The irrational has a place only where infinite divisibility is possible (ibid., p.48).

Thus Proclus believes that in order to have incommensurables one needs infinite divisibility.   In fact, as we saw in the last chapter, to have the *existence* of one incommensurable magnitude (for example, $\sqrt{2}$) this is not necessary.   For what the proof in Euclid book X, Proposition 117 shows is that $\sqrt{2}$ cannot be measured exactly.

There is a hidden assumption which is known as Archimedes' axiom.   We present this in modern language in terms of lengths.

Given any two (finite lengths   A   and   B   then there is a natural number   n   such that   nB > A, i.e.   n   lengths   B   laid out successively are longer than   A.

(In this case   n = 4.)

Our procedure in determining the decimal value of a given length   A   when   B   is the unit length consists of the following algorithm:

Stage 1.   Write down   n   for the least   n   such that   nB > A   and then write a decimal point.   Cut off   (n-1)B   from   A   and call the result A_1.

Stage k+1.   Let   B′   be one-tenth the length of the   B   used in stage k.   Find the least   n   such that   nB′ > A_k where A_k is the length of   A   left after stage k.   Write down   (n-1).   Cut off   (n-1)B′   from A_k and call the result A_{k+1}.   If A_{k+1} is not of zero length go to stage k+2.

What is written down is the usual decimal number for the length   A   measured in terms of the unit length   B.   Clearly Archimedes' axiom is used at each stage.[5]

Now taking one tenth of a given length is something Euclid could easily do.   Suppose we want to find a tenth of the length   B.

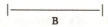

Take any length   C, draw   10C, join the ends by a line as shown below, and then draw a line parallel to it (shown dotted) which cuts off the required tenth of   B.

135

In this way it is easy to see how Proclus saw infinity coming in with incommensurables. If one has $\sqrt{2}$ which has no common measure with the unit, then the algorithm can be applied forever; moreover, it seems apparent that different decimals will give different lengths. This is neither true nor obvious in fact! That it is not true follows from the fact that 1.999... $\pm$ 2.000 ... . That it is not obvious was pointed out by Cantor (see below §6).

3. The Renaissance and after

The discovery of incommensurables or irrational magnitudes led to a great deal of interest in, and work on, such magnitudes. Whether one takes the widely held view that the discovery of irrationals provoked a crisis in Greek mathematics or the newer view of Knorr [1975], that the study of irrationals was an important focus of work in Greek mathematics, it is clear that irrationals were a continual source of concern and difficulty for a very long time. As we have seen already in the *Theaetetus*, the irrationality of $\sqrt{3},...,\sqrt{17}$ (omitting, of course, $\sqrt{4}$, $\sqrt{9}$ and $\sqrt{16}$) was known early (see for example Knorr [1975], p.21-61). But this interest was concentrated in the classification of constructible magnitudes. For the ancient Greeks this meant construction by ruler and compass and therefore was limited to those which are equivalent to numbers formed by using addition, subtraction, multiplication and division and the taking of square roots. With, of course, the added proviso that all should be positive. Thus in Euclid's book X we have a tremendous concern with (some of) these ratios of magnitudes.

In the East, however, there was no concern with proof and the logical problems that incommensurables caused. The Chinese, for example, simply measured lengths to an approximate degree of precision. However, the Chinese used special terms to denote each decimal place (see, for example, Li Yan and Du Shiran [1987], p.83). There was always an identification of lengths and numbers rather than a philosophical gap between them. This is because the mathematics of the Chinese was practical and algorithmic (see Li Wenlin [preprint]).

In Europe, however, Euclid's attitude and the concept of proof dominated thinking about magnitudes and numbers even up to the sixteenth century. For the Arabs, as we have seen in chapter III, the situation was somewhat different, for the solution of cubic equations by geometric methods meant that lines were being constructed whose ratios to given lines involved cube roots.

As we have noted in chapter IV, by the end of the fifteenth century European mathematicians possessed considerable expertise in the employment not only of cube roots but also of many other kinds of radical expressions. Here by "radical expression", we mean an expression involving plus, minus, times, divided by, and the extraction of n-th roots for particular numbers (i.e. positive integers) n.

136

When we turn to the work of the German algebraist Stifel (*ca.* 1487-1567), we find that he believed that *all* irrational ratios corresponded to radical expressions. Book II (p.103r) of his *Arithmetica Integra* (Stifel [1544]) of 1544 is entirely devoted to irrationals.

> *It is rightly disputed of irrational numbers whether they are true numbers or false. For because in proving with geometric figures, where rational numbers desert us, irrationals take their place, and they prove precisely what rational numbers were not able to prove, at least from the demonstrations which we know of: we are moved and compelled to admit they are correct, as is clear from their effects, which we feel to be real, certain and constant.*[6]

Now he distinguishes sharply between lengths and numbers and says that lengths cannot be precisely captured by numbers. For he continues:

> *But other things induce us to a different assertion, so that we are clearly compelled to deny that irrational numbers are numbers. In fact, where we try to subject them to numeration, and to be proportional to rational numbers, we find they continually escape us, so that none of them can be apprehended precisely as its proper self ... But it is not possible to call a number real which is of such a kind that it is devoid of precision, and has no known proportion to real numbers. Therefore, just as an infinite number is not a number, so too an irrational number is not a real number, because it lies under some cloud of infinity. And the ratio of an irrational number to a rational number will be no less uncertain than that of an infinite to a finite (ibid., p.103).*[7]

So if irrationals were true numbers they would be integers or fractions. But it is easy to show that they are not.

Shortly afterwards (*ibid.*, p.104r), he claims Euclid denied irrational numbers were numbers in Proposition X.5 (see above).

This done he goes on to classify irrationals following Euclid at first and then those involving not only square roots but also cube roots, fourth roots, fifth roots and so on. Then he deals with composites (for example $\sqrt[3]{12} + \sqrt[3]{6}$, p. 109v), finally ending up with expressions as complicated as

$$\sqrt[3]{\{(\sqrt{20} - \sqrt{10}) + \sqrt{(\sqrt{8} - \sqrt{2})}\}} \qquad \text{(p.111r)}.$$

Maurolico (whom we met in chapter II), writing in 1557, has a similar approach. Part two of his second book ([1575], p.127 f.) is dedicated to irrational quantities and, in the same way as he began Book I part I, he presents a list of types of irrational expression (p.128-130) and on p.131 gives a table which starts off

Rationalis ...

Quantitas {

Irrationalis { . . .
                . . .

The upper branch from *Irrationalis* splits into simple expressions leading to square, cube, fourth power, etc., and the lower goes eventually on the one hand to the binomia[8] in the style of Euclid (Book X) and on the other to sums and products employing medials[9]. He starts off from a unit line and then constructs all the others (see p.85).

Thus Maurolico does not consider magnitudes which cannot be described by radical expressions. The nearest he gets to considering the circle is when he is looking at the side of a regular polygon (triangle, square, pentagon and some other polygons) inscribed in a circle of rational diameter.

Stifel on the other hand in [1544] devotes an appendix to Book II (p.224r-226r) to the quadrature of the circle. Here he distinguishes firmly between physical circles (*circulus physicus*) and mathematical circles (*circulus mathematicus*). The former is something in the world and one can measure and calculate its area.

> 3. *The physical circle is a particular image of the mathematical circle.*

He then goes on to describe the mathematical circle as the last polygon.

> 4. *The triangle is the first of all the polygons.*
> 5. *The last of all the polygons is the circle.*
> 6. *Therefore the mathematical circle is rightly described as the polygon of infinitely many sides.*
> 7. *And thus the circumference of the mathematical circle receives no number, neither rational nor irrational.*[10]

And he concludes (on p.226r) that just because one can give an upper and a lower estimate for the quadrature of the circle, it does *not* follow that one can give an exact number.

His logic is faulty by nineteenth or twentieth century standards but his perception in anticipating that there are lengths (and areas) which cannot be given a description in terms of radicals (and +, −, etc.) is remarkable.

Given the approach of Maurolico (and also of Stifel), it is easier to see the importance that the quadrature of the circle and the finding of the length of the circumference assumed. Stifel, as we have seen, did not believe that the circumference had a length, though he knew that one could approximate it as closely as required.

Barrow, writing a hundred years later, thought of magnitudes as given by constructions (Barrow [1860], lecture IV, p.72). And when he considered the circumference of a circle he opined that it could be regarded as known but could not be compared with a straight line.

That is, the circle itself, seems to have a perimeter, with an infinite number of steps, incommensurable with the radius (Lect. XV, p.249).[11]

Barrow felt that magnitudes were primary (*ibid.*, lect. XII, p.187) and then, corresponding to them, there were, but only in some cases, numbers. Thus he defines equality of numbers in terms of magnitudes (*ibid.*, lect. XII, p.193).

Finally equal numbers can be defined as those things which magnitudes, divided in the same way, represent (or more accurately, divided in any other similar way whatsoever).[12]

All this is despite the fact that he regards all quantity as continuous (*ibid,.* lect. II, p.39).

Thus we can see lines as having certain points on them constructible by using geometric methods and even by taking arbitrary n-th roots (where n is a positive integer). There was a tendency to regard these points as being *all* the points on the line, though even as far back as the thirteenth century we have Fibonacci proving that the solution of a certain cubic equation was not one of Euclid's irrationals (see above, chapter III). Of course the number that Fibonacci approximated was an algebraic number.[13]

The awareness of the difference between numbers and magnitudes is nicely brought out by John Dee in his preface to Billingsley's edition of Euclid, the first edition in English.

But, as (by degrees) Number did come to our perceiverance: So, by visible formes, we are holpen to imagine, what our Line Mathematicall, is (Billingsley [1570], Preface aiir).

And a little later (on *.iir) he writes:

Arithmetike of Radicall numbers: *Not, of* **Irrationall** *or* **Surd Numbers:**

Practise hath led **Numbers** *farder, and hath framed them, to take upon them, the shew of* **Magnitudes** *propertie: Which is* **Incommensurabilitie** *and* **Irrationalitie:** *(For in pure* **Arithmetike,** *an* **Unit,** *is the common Measure of all Numbers.) And here Numbers are become, as Lynes, Playnes and Solides: some tymes* **Rationall,** *some tymes* **Irrationall.** *And have proper and peculier characters (as* √⅜. √&.*) and so of other ... Wherfore the practiser estemeth this, a diverse* **Arithmetike** *from the other.*

139

4. Infinite expressions

The Arabs had considered finite decimals at an early date. Thus Saidan [1966,1978] has published work of al-Uqlidisi from 952/3 A.D. in which decimal fractions are calculated to as great a (finite) degree of precision as required (Saidan [1978], p.110 and p.481). Later al-Kāshī, who died in 1436/7 A.D., also used decimal fractions (Saidan, *op.cit.*, p.481). The work of these two authors did not become known in the West before the time of Stevin.

Now although the Arab authors used as many decimal places as they required, it was Stevin's notational system which made it explicit that decimals could go on forever even though Stevin did not consider infinite decimals.

Stevin introduced his decimal notation in *De Thiende* (Stevin [1585]). This work was very short, but had great influence. He proceeded by giving simple examples. Thus treating of what we write as .3759 he said (Stevin [1958], p.407)[14]:

As 3 ①️ 7 ②️ 5 ③️ 9 ④️*, that is to say:* 3 **primes,** 7 **seconds,** 5 **thirds,** 9 **fourths,** *and so proceeding infinitely, but to speak of their value, you may note that according to this definition the said numbers are*

$$\frac{3}{10} \ , \ \frac{7}{100} \ , \ \frac{5}{1000} \ , \ \frac{9}{10000} \ , \ together \ \frac{3759}{10000} \ .$$

Here we see that he recognizes that there are an infinite number of decimal places but he does not quite include the idea of infinite decimal. Only finite approximations are considered.

In his *Arithmétique*, Stevin [1635] points out that all is easy if one knows how to handle radicals and he says he will show how to deal with all kinds of binomia. On p. 215 Stevin quotes Euclid X.2 and notes that if the Euclidean algorithm does not terminate then the lengths are incommensurable.

But although this theorem is valid, nevertheless we cannot recognize by such experience the incommensurability of two given magnitudes: Firstly because of the error of our eyes and hands (which cannot perfectly see and divide) we would judge in the end, that all magnitudes, incommensurable as well as commensurate, are commensurable. Second, even though it were possible for us to subtract by due process, several hundred thousand times, the smaller magnitude from the larger, and continue that for several thousands of years, nevertheless (the two given numbers being incommensurable) one would labour eternally, always remaining ignorant of what could still happen at the end; This manner of cognition is therefore not legitimate, but rather an impossible position, so that in this way nothing may be stated about what really happens in nature; so this incommensurability is only to be noted for incommensurable numbers; this Euclid knew very well, and further, that the finding of incommenssurability was not sufficient for the following propositions (because his tenth proposition shows how to find incommensurable magnitudes by means of numbers), and so he explained this legitimately in the 8th proposition using

*numbers, and so shall we in this first part, [which deals with] numbers, as
follows.*

Definition I.

*Incommensurable magnitudes are those, whose corresponding numbers are
incommensurable.*[15]

And then, when describing the extraction of cube roots, he says:

*So that one can thus infinitely [closely] approach the true value but
never can one by such means arrive at [the true value]; the reason for which
is (...) that the incommensurable cannot be commensurable (ibid., p.126).*[165]

But Stevin was quite explicitly committed to regarding all the various
kinds of real number as belonging to the same family (see above, Chapter IV,
§3). At the end of his *Arithmétique* [1635] he lists his main theses, of
which the first four are:

Thesis I

That unity is a number.

Thesis II

*That any given numbers can be
square, cubes, fourth powers, &c.*

Thesis III

That any given root is a number.

Thesis IV

*That there are no absurd, irrational, irregular,
inexplicable or surd numbers (ibid., p.218).*[17,18,19]

In England there was a gradual move towards admitting numbers
corresponding to infinite computations. Thus Oughtred's *Clavis
Mathematicae* (Oughtred [1648]) of 1648 shows a slight inclination that way,
but he only considers numbers given by proportionalities and surds.

Wallis did fully accept such computations. Early in his *Algebra*
(Wallis [1684/5]), chapter LXXXVII, he is guarded about the use of decimals.

*Not that the value of all Fractions Whatsoever may ... be accurately
expressed [in decimal notation] ... but because many times it may be done
exactly ... and when it cannot be done exactly, yet ... we may come very
near the truth, to what degree of approximation we please ... .*

He then gives examples of calculating
12 1/3 + 10 1/2 + 7/25 + 3 5/11 + $\sqrt{2}$ + $\sqrt{3}$ - $\sqrt{2}$ to prove his point. Later
in the work he is concerned with his form of induction (see above, chapter
II, §. 5) and this leads him to consider expressing proportions not hitherto
possessing any notation. Discussing these, he writes:

But the Result is, that such Proportion is not to be expressed in the commonly received ways of Notation: And particularly, that for the Circles Quadrature (Wallis [1684/5], p.315).

Then he goes on to say why one cannot use integers or surds or powers and after this to consider approximations.

Now (as in other Incommensurable Quantities), though the Proportion cannot be accurately expressed in absolute Numbers: Yet by continued Approximation it may; so as to approach nearer to it, than any difference assignable (ibid., p.317).

After this he gives various types of approximation including continued fractions and successive square roots (*ibid.*, p.317-322). This all prepares the way for his discussion of infinite series (*ibid.*, p.324, ch. LXXXVIII).

However, seventeenth century writers on series did not worry about convergence and there are some curious "results".[20] The study of series developed substantially and in the eighteenth century Waring clearly defined convergence and even showed that $1/1^m + 1/2^m + 1/3^m + ...$ is convergent if $m > 1$ and divergent when $m < 1$. His definition of convergence is as follows:

Def. 1. Given an infinite series $a + b + c + d + e + f + g + h + \&c.$ and if the successive sums a, $a + b$, $a + b + c$, $a + b + c + d$, &c. continually converge to the sum of the series, and finally approach nearer than any given difference, then it can be said that this series is convergent (Waring [1776], p.294).[21]

As Kline ([1972], p.593) remarks, nothing was done in the eighteenth century to clarify the notion of irrational number. However, there were many approaches in the nineteenth century which led to equivalent descriptions. Those are almost the standard treatments we have today. We shall not treat all of them.

5. The Nineteenth Century

The way to a new description of the numbers associated with the line and of numbers not corresponding to radical expressions was opened by the demonstrations that for example e, the basis of the natural logarithms, and e^2 are irrational (Euler), that π is irrational (Lambert) and that transcendental numbers exist (Liouville).[22] Also the new work on functions and Fourier series showed the need for a deeper study (cf. Kline [1972], p.979, 949).

Kline remarks:

Curiously enough, the erection of a theory of irrationals required not much more than a new point of view. Euclid in Book V of the Elements had

*treated incommensurable ratios of magnitudes and had defined the equality
and inequality of such ratios (ibid., p.982).*

But we shall show that one thing that was required was a new axiom – a
major assumption.

The earliest attempt to give a theory of what we now call real numbers
was made by Hamilton.

*(§26) The existence of these incommensurables [such as $\sqrt{2}$, $\sqrt{3}$, $\sqrt{5}$,...]
(or the necessity of conceiving them to exist,) is so curious and remarkable
a result, that it may be usefully confirmed by an additional proof of the
general existence of square-roots of positive ratios, which will also offer
an opportunity of considering some other important principles (Hamilton
[1967], p.54).*

He then went on to consider a ratio *a*   satisfying

$$a > a' \;,\; a > b' \;,\; a > c' \;,\ldots \\ a < a'' \;,\; a < b'' \;,\; a < c'' \;,\ldots \left.\right\}$$

(ibid., p.55) where each of   *a'*, *b'*, *c'*,... is less than   *a''*, *b''*, *c''*,...
This is, of course, very close to the Eudoxan approach to ratio (see for
example Eudoxus in [a] and Euclid Book V, Definition 5).   After this he
considered ratios given by a law, pointing out that unless the conditions
were incompatible, "they [could] never exclude *all ratios*   *a*   *whatever*"
(ibid., p.56).   In defining a unique ratio, as for example, $\sqrt{2}$, he uses his
idea of the continuous and constant increase of a ratio   *x*   from zero so
that its square x^2 will

*... pass successively but only once through every state of positive
ratio* ♭*: and therefore that every determined positive ratio* ♭ *has one
determined positive square root* $\sqrt{♭}$*, which will be commensurable or
incommensurable, according as* ♭ *can or cannot be expressed as the square
of a fraction. When* ♭ *cannot be so expressed, it is still possible to*
approximate in fractions *to the incommensurable square root* $\sqrt{♭}$*, by choosing
successively larger and larger positive denominators ... (ibid., p.59).*

He then proceeded (p.60-74) to employ the same technique for dealing
with b^{α} for any number   α   commensurable or incommensurable.

The assumption he has not made explicit is that it *is* possible for a
quantity to increase constantly and continuously.   Indeed, here is where an
extra axiom is needed.   By slightly modifying his approach one could have
defined a real number   α   (or, in his terminology, a commensurable or
incommensurable quantity) *to be* the collection of ratios

$$a > a' \;,\; a > b' \;,\; a > c' \;,\; \ldots \\ a < a'' \;,\; a < b'' \;,\; a < c'' \;,\ldots \left.\right\}$$

The fact that all such "real numbers" could then be compared would,
however, have required a considerable amount of technical argument and
development to show, for example, that the real numbers can be ordered and

143

that all are comparable (under the obvious definition, or should we say extension, of <).

Because Hamilton insisted on a law for defining his sets of ratios which would determine a real number, it is not clear that Hamilton in fact intended or believed that he was treating *all* real numbers. (Kline describes Hamilton's work as unfinished.)

The problem of treating continuously variable quantities had been attacked by Descartes. His use of letters in what we now call Cartesian or co-ordinate geometry allowed him to talk of operations on varying lengths. For example, a/b is determined by the geometric solution of the proportionality

$$a : b = x : 1.$$

a is to b as a/b is to 1 (Descartes [1954], p.3). This construction is in Euclid. It is clear that this works for all lengths a and b. However, it was not until the second half of the nineteenth century that a satisfactory theory of real (including irrational) numbers developed.

The Greeks had "deftly sidestepped" the problem of incommensurables by substituting magnitude for number (Wilder [1973], p.111). The nineteenth century growth of analysis and the study of differential equations demanded a solution.

Ch. Méray was the first[23] to find[24] a purely arithmetic meaning for the expression **irrational number** *(Molk [1904], p.148)*[25,26]

Méray *defines* real numbers as equivalence classes of Cauchy sequences; at least that is the modern descriptions. He uses sequences of the form $\nu_{m,n...}$, permitting several indices unlike our present practice of using only one index. He says:

...the necessary and sufficient condition for a given variable $\nu_{m,n}$ to tend to a commensurable or incommensurable limit is that it be convergent, that is to say, that the difference

$$\nu_{m'',n'',...} - \nu_{m',n',...}$$

tend to zero as the integers m', n',...,m'', n'',... all increase indefinitely in some way (Méray [1894], p.36).[27]

What Méray does is to define two sequences $u_1,u_2,...$ and $v_1,v_2,...$ to be *equivalent* if $u_m - v_n$ tends to zero as m and n tend to infinity. (We use a single subscript for clarity.) He then introduces "fictitious numbers" (*nombres fictifs*) corresponding to such sequences and *defines* two fictitious numbers to be equal if the sequences are equivalent in the sense defined above. It is then of course easy to check, as one does in a basic analysis course, that these fictitious numbers obey the same laws as rational numbers for addition, multiplication, etc. and that a rational number, a, can be identified with the fictitious number given by the constant sequence a, a, a, ...

In this way rational numbers can be regarded as real numbers (technically, they can be embedded in the real numbers). However, by admitting "fictitious" numbers by fiat, Méray is tacitly assuming the axiom:

every convergent sequence has a [real number as its] limit.

That he had some idea that he could not *prove* all real numbers exist seems clear from his use of the name "fictitious number".

In the same year that Méray published his paper, Heine also introduced a definition of real number. Others who developed definitions are du Bois-Reymond (see [Die allgemeine Funktionentheorie, vol. 1, Tübingen 1882]), Cantor and Dedekind. (For the last two of these authors see below.)

du Bois-Reymond took the view that there was a psychological necessity for the existence of real numbers because of the connexion between numbers and measurable magnitudes. He felt that it was going against the history of mathematics to separate the study of number from that of magnitude and therefore that the theory of real numbers was justified by its applications. (See Molk [1904], p.155 f.) Certainly many mathematicians behave in this way and there have been no failures of the theory so far and none are expected. It does, however, appear that even for du Bois-Reymond non-mathematical features such as the history of actual applications must intrude.

du Bois-Reymond depended on applications, Hamilton used time as a basis. Thus Kline writes:

[Hamilton] based his notion of all the numbers, rationals and irrationals, on time, an unsatisfactory basis for mathematics (though regarded by many, following Kant, as a basic intuition) (Kline [1972], p.983).

6. Cantor

How nicely this echoes Brouwer's criticism of Dedekind (see chapter II, §. 6) for using the realm of his thoughts for a proof of the existence of an infinite set. But this is an area in which theology abounds. Cantor made this clear in his series of articles [1879-1884] (see Cantor [1932], where speaking of the line as continuum, he wrote:

In the development of science the concept of the "continuum" has in general not only played a significant rôle, but also continuously evoked the greatest differences of opinion and even violent disputes. The cause of this is perhaps the fact that its underlying notion as a phenomenon has taken on a fundamentally different content with the dissenters because the precise and complete definition of the concept had not been transmitted to them; but also perhaps, and this seems the most likely to me, those Greeks who may have grasped the notion of the continuum first, even then did not work it out with the clarity and completeness which would have been necessary to exclude the possibility of different opinions (conceptions) among their successors. Thus we see that **Leucippus, Democritus** *and* **Aristotle** *regarded the continuum as a composite which consists of parts divisible without end, whereas* **Epicurus** *and* **Lucretius** *construct the same from its atoms as finite things; subsequently this caused a great controversy among the philosophers, of whom some followed* **Aristotle**, *others* **Epicurus**, *others again, to keep away from this controversy, stated with* **Thomas Aquinas**[28] *that the continuum consisted neither of infinitely many, nor of a finite number of parts, but of no parts at all; this latter view appears to me to be less a statement of fact than the silent admission that one has not got to the root of the matter and prefers to elegantly side-step it.*

Here we see the **mediaeval scholastic origin** *of an opinion which we still find being held today, that the continuum is an irreducible concept or even, as others express it, a purely* **a priori** *intuition which hardly permits specification by concepts; any attempt to define this* **mystery** *arithmetically is viewed as an inadmissible interference and rejected with appropriate force; timid natures thereby receive the impression that with the continuum it is not a question of a* **mathematical-logical** *concept but, rather, a* **religious dogma** *which is being dealt with. Far be it from me to again conjure up this controversy; moreover, lack of space prevents a closer discussion of the subject in this limited framework; I only feel compelled here to develop the concept of the continuum as briefly as possible, taking into account also the* **mathematical** *theory of numbers, by looking at it logically and soberly and with a view to its need in the theory of sets (Mannigfaltigkeitslehre). This treatment has not been easy for me because among those mathematicians on whose authority I prefer to rely, not a single one has dealt with the continuum more precisely in the sense I need here (Cantor [1932], p.190-191).*[29]

Cantor first introduced his treatment of real numbers in a paper of 1872. He knew of Méray's work (see Molk [1904], p.152). He assumes the rational numbers as a foundation (Cantor [1932], p.92). Then, like Méray, he considers sequences of rationals

$$a_1, \ a_2, \ldots, \ a_n, \ldots \qquad (1)$$

such that

the difference $a_{n+m} - a_n$ *becomes infinitely small with increasing* n, *whatever the positive integer* m *may be, or in other words, that for an arbitrary assumed (positive, rational)* ϵ *there is an integer* n_1 *such that* $|a_{n+m} - a_n| < \epsilon$ *if* $n \geq n_1$ *and* m *is an arbitrary positive integer. I express this quality of the sequence (1) in words as "The sequence (1) has a* **determinate limit b**" *(Cantor [1932], p.93).*[30]

Here b is specifically given by the sequence (1). He goes on to define equality and inequalities between such symbols b, b' with associated sequences $a_1, \ a_2, \ldots$ and $a'_1, \ a'_2, \ldots$ by setting, for example, b

$< b'$ if $a'_n - a'_n > \epsilon$ for some fixed positive rational ϵ and all sufficiently large n. From this he proceeds to the elementary operations defining $b + b'$ in terms of the sequence with elements $a_n + a'_n$.

Now Méray had been content to take only rational numbers as elements of his sequences. Cantor goes further and considers the effect of repeating the process on what are initially Méray's fictitious numbers.

Cantor considers (op.cit., p.94) sequences b_1, b_2,..., b_n,... using his newly acquired meaning for $b_{n+m} - b_n$ which is given by using the associated α's in the sequences determining the b's. The process is repeated to give new numbers (Zahlengrösse) of levels 1, 2, 3,... (where the b's are in level 1).

To every rational number a there corresponds a b of level 1 (by considering the constant sequence $a, a, a, \ldots$) but not conversely, while in fact for higher levels there is a correspondence, though Cantor did not worry about this. He later wrote (see Cantor [1932], p.188, §III.4):

... it was not far from my mind to introduce, through the basic sequences of second, third order, etc., new numbers which would not already be determinable (definable) through the basic sequences of the first order, but I had only the abstractly different form of the given one in mind; this is clearly evident from individual passages in my work itself.[31]

He then has a most perceptive passage. He considers the line with directions $+$ and $-$, a fixed point o and a measure of distance:

If this distance has a rational relation to the unit of measure, then it is expressed by a numerical quantity in the domain A [the rationals]; otherwise, if the point is one known through a construction, it is always possible to give a sequence

$$a_1, a_2, \ldots a_n \ldots \qquad (1)$$

which has the property indicated in §1 [as above] and relates to the distance in question in such a way that the points on the straight line, to which the distances, $a_1, a_2, \ldots a_n, \ldots$ are assigned, approach in infinity the point to be determined with increasing n. This we express by saying: the distance of the point to be determined from point o is equal to b, where b is the numerical quantity corresponding to sequence (1) ... [And similarly for higher order numerical quantity.] In order to complete the connexion presented in this § of the domains of the quantities defined in §1 with the geometry of the straight line, one must only add an axiom which simply says that inversely every numerical quantity also has a determined point on the straight line, whose coordinate is equal to that quantity, indeed, equal in the sense in which this is explained in this §. I call this proposition an axiom because by its nature it cannot be universally proved. A certain objectivity is then subsequently gained thereby for the quantities, although they are quite independent of this (objectivity) (p.96-97).[32,33]

Thus he clearly says that one needs an *axiom* which says that for each real number there is a corresponding point on the line.[34,35]

Cantor's awareness of the need for the extra axiom was shared by Stolz, but this was perhaps for a different reason.

The hypothesis in the fundamental theorem of the exhaustion method, that any two geometric quantities of the same type are comparable is, theoretically at any rate, inadmissible; for analytic geometry can establish curves which have no finite length.[36] Here there is no other alternative than to define the lengths of the curves, the areas and the volumes as limits (Stolz [1885], p.78).[37]

Stolz also considers infinite decimals, regarding them as the limits of finite approximations. In particular he shows (*ibid.*, p.101) that every non-periodic decimal is irrational. He regards these as giving new objects, different from the rational numbers.

... thus let us imagine a new object, different from every rational number, produced by the unlimited sequence ϕ_0, ϕ_1,... (ibid., p.105).[38]

Thus by this stage we have moved to the situation where the emphasis is no longer on geometrical constructions, but depends entirely on numbers (and sequences of numbers).

7. Dedekind

Dedekind's approach depended, too, on the creation of new numbers from (sets of) rational numbers. Although his work *Stetigkeit und Irrationalzahlen* (see Dedekind [1968]) did not appear in print until 1872, he notes in the preface that he originally had the ideas in 1858.

Since every rational number divides the rational numbers into two classes: those which are smaller and those that are not[39], Dedekind extended this notion and identified the pairs of classes with numbers. In the case of irrational numbers such as $\sqrt{2}$ the rational numbers are still divided into two classes in the same way. He then took this as his definition of an arbitrary real number.

If all points of the straight line fall into two classes such that every point of the first class lies to the left of every point of the second class, then there exists one and only one point which produces this division of all points into two classes, this severing of the straight line into two portions (English translation from Beman [1963], p.11).

Here again Cantor's axiom is implicit. Dedekind's definition has two aspects. First of all he defines pairs of classes (A,B) where A is the first class and B the second in the above definition. This is equivalent to Méray's fictitious numbers. Secondly, Cantor's axiom then yields a point which produces the division into two classes.[40]

This unique point is the particular real number considered.

Working from Dedekind's definition, one can then prove the

Theorem. *If A is any set of* **real** *numbers such that every real number in A is less than some real number b then there is a (unique) least real number* b_0 *greater than or equal to every real number in A.*

Dedekind's real numbers had sufficient properties for the adequate development of the foundations of the calculus. His formulation and its equivalents have become accepted and now, on opening a book on analysis, one usually finds a description of "the real line" which is entirely divorced from the geometric viewpoint of earlier centuries. Nowadays the "real line" is considered as completely arithmetized; in effect, Cantor's axiom has been replaced by a *definition* : the real line = $\mathbb{R}$, the set of real numbers.

8. Formalization

The identification of "real line" with the geometric line appears to have obscured the fact that, although one now has a collection of axioms which accord with an intuitive view of the geometric line, nevertheless there is no guarantee that one has captured the full description of the geometric line.[41]

Hilbert, in giving axioms for *geometry* in fact gave a set of axioms which characterize the real numbers. One crucial innovation of his was the completeness axiom. This says, simply enough, that the points on a line are to comprise a maximal set satisfying the other axioms (Hilbert [1971], p.26). The completeness axiom has a different character from all the other axioms (which include, for example, the axiom that between any two distinct points there is a third point on the line joining them). It refers to the totality of points. Previous authors, such as Stifel, Maurolico, Leibniz, had only considered ways of producing more real numbers, for example, by taking n-th roots. They never considered the possibility that the generation of more numbers or points, which is clearly an infinitely proceeding process, could be completed. The idea of a totality which was, as we say today, closed under certain operations was not considered as such. With the growth of abstract algebra and set theory in the nineteenth century such ideas came in.

The completeness axiom guarantees (in this particular case) the uniqueness of the set of real numbers where by "uniqueness" we mean that any other presentation of a set satisfying the same axioms is essentially the same. In mathematical terminology, it is isomorphic to the real numbers.[42]

It would seem that with Hilbert's axioms the nature of the real numbers has been made totally clear and precise. In fact this is not the case. Because of the need to consider the totality of real numbers, Hilbert's axioms lead us into infinite set theory. There we encounter a number of difficulties and, in the early development of set theory, paradoxes which

need clarification.   The process of clarification is still continuing.[43]
Our concern is with number and we shall not pursue the set theory here.

In an effort to avoid the paradoxes of set theory, Hilbert formalized mathematics.   Unfortunately, in the process of formalization the uniqueness of the real numbers is lost.[44]   If we take the formalized axioms for the real numbers then we do have the real numbers as one interpretation, or model, of the formalized axioms.   However there are many other possible interpretations which are not at all the same as the real numbers, that is, they are not isomorphic to the real numbers.   These interpretations are the so-called "non-standard models of analysis".   They were first introduced in the middle sixties by Abraham Robinson [1966] though the technique he used had been known since 1915 (see Skolem [1934]).

The nineteenth century treatment of the real numbers had resolved difficulties which had arisen in calculus through the use of infinitesimal quantities.   The theories of Cantor and Dedekind allowed the development of the calculus without infinitesimals.   On the other hand one of the most interesting aspects of non-standard models of analysis is that they contain infinitesimals in a totally coherent and consistent way.   However they do violate one axiom.

In the comments following the introduction of his completeness axiom, Hilbert [1971], p. 26) explicitly remarks that Archimedes' axiom must be included here.   In numerical terms the Archimedean axiom says that given any positive numbers a, b then if a be added to itself over and over again then after a *finite* number of additions the sum will be greater than b. That is, for all a, b there exists an *integer* n such that a + ... + a (n times) is greater than b.   This axiom is in Euclid as definition V.5.

Robinson's non-standard models of analysis violate Archimedes' axiom while providing interpretations in which all the assertions about the ordinary real numbers in the ordinary language of mathematics[45] are true just in case they are true in the ordinary real numbers.   However, in the non-standard models there are infinitesimals, that is to say "numbers" bigger than zero but smaller than any ordinary fraction 1/n.   It follows that any finite multiple of an infinitesimal is also less than any positive ordinary fraction.   This clearly contradicts Archimedes' axiom. Nevertheless results proved using infinitesimals which are written in the formal language are valid for the standard real numbers.   Thus Abraham Robinson's non-standard models give us alternative, and very often simpler proofs of results in calculus and analysis.[46]   From the point of view of the theorems of the formal language it does not matter whether one works with non-standard real numbers or with the standard ones.   Further there is no way of verifying what is the "real" (in the sense of genuine) structure of the geometric line.   All the work we have discussed has been done in the context of numbers rather than that of geometry.   It leaves the question of the distribution of points on the geometric line completely untouched.   This is how the situation stands at present.   There is a system of numbers which works very well indeed for calculus and analysis. Everyone regards this system of numbers, in actual practice, as being identifiable with the points on the line and Cantor's explicit introduction of an axiom to justify that identification is generally forgotten.

9. Conclusion

As we have seen, Egyptian mathematics (and Babylonian, too, for that matter) was very much concerned with geometry in an arithmetic framework, that is to say, with problems of mensuration. The time of Thales is generally regarded as heralding the start of Euclidean-style geometry, though it is by no means clear how close to the presentation in the books of Euclid the studies were in Thales' time. Further, in Babylon and Egypt (as is obvious from the problems considered on many cuneiform tablets and in the Rhind Papyrus) and presumably in Greece too, natural numbers were being used in commercial transactions and in particular the dimensions of land were being measured. Thus there was an application of (natural) numbers to concrete geometric geometric figures (fields, etc.). So long as there is only this one-way application the correlation works well. Once the idea emerged that the converse correspondence should also work, a whole new part of mathematics was opened up. This correspondence asserts that to each (even merely straight) edge of, say, a field there corresponds a number (of some kind). So long as it is possible (by, for example, shortening the basic unit of length) to measure each line as an integral multiple of the basic length, there is no need to extend the number beyond the natural numbers. It is, in fact, not even necessary to employ fractions explicitly. One can simply take a new unit which is, let us say one tenth of the previous unit in length. When it is insisted that just as numbers may be applied to lengths, so too must every length have a number corresponding to it, we arrive at the point where the possibility of incommensurables arises.[47]

Now there is, in fact, no *necessity* to make the connexion (between numbers and lengths) in *both* directions.[48]

According to Aristotle, the Pythagoreans regarded numbers (meaning natural numbers rather than fractions or other kinds of number) as being "the first things in the whole of nature". But by Aristotle's time the interconnexion between lines and numbers had been hypostasized. Aristotle saw the difficulty in the Pythagorean view that all things are number but had already opted for the view which regards real space as continuous.

For it is not true to speak of indivisible spatial magnitudes; and however much there might be magnitudes of this sort, units at least have not magnitude; and how can a magnitude be composed of indivisibles? But arithmetical number, at least, consists of units, while these thinkers identify number with real things; at any rate they apply their propositions to bodies as if they consisted of those numbers (Metaphysics 986a).

Considerably later Proclus (5th century A.D.), writing on Euclid, also points out how the existence of irrationals is what distinguishes geometry and arithmetic.

If there were no infinity, all magnitudes would be commensurable and there would be nothing inexpressible or irrational, features that are thought to distinguish geometry from arithmetic; ... (Proclus [1970], p.5).

Thus this later view allowed the division of mathematics into arithmetic and geometry, while maintaining there was a close connexion. Gow goes so far as to state:

> **Number absolute** *was the field of arithmetic:* number applied *of music:* **stationary magnitude** *of geometry,* **magnitude in motion** *of spheric or astronomy.* *But they did not so strictly dissociate discrete from continuous quantity.* *An arithmetical fact had its analogue in geometry and* **vice versa.** *Pythagorean arithmetic and geometry should therefore be treated together, ... (Gow [1884], p.71-2).*

Now a most important question is how soon the "*vice versa*" analogue of geometry in arithmetic was appreciated. I suggest it was closely associated with the discovery of irrational numbers for it is then that the question becomes significant.

At this discovery the foundation of the Pythagorean philosophy, namely that the whole essence of the world is number, meaning natural number, is destroyed. I therefore suggest that when we read in the Scholium to book X of Euclid that "*if any psyche were to make an assault on such a form of existence as this, he would simply make obvious and clear the fact that he is being carried under into the sea of creation and is being overwhelmed by the unstable surges of it*", it merely means what it says. I suggest that the discovery had a profoundly disturbing effect on the consciousness of the discoverer (be it Hippasus, as suggested by von Fritz [1945], or someone else). Moreover, to overthrow the accepted doctrine by means which were acceptable to the givers of the doctrine must have made the effect all the more profound and disturbing. This view is even consistent with that of Knorr [1975] for he claims that the discovery was a great stimulus to the study of mathematics. Now, surely, in the discovery of incommensurables, some proof[49] must have been necessary, though of course it might have borne the merest shadow of resemblance to our sort of proof. Even if the 'proof' consisted of demonstrating, say using pebbles or dots, that the ratio of the diagonal of a square (or the ratio of a pentagon or pentagram's diagonal) so its side was incommensurable, the result was of such innovative importance that it generated a legend of quite dramatic nature and gave a great stimulus to the study of mathematics and in particular incommensurable magnitudes.

Subsequent developments consisted in a deep study of the nature of these incommensurable magnitudes but there was also a profound shift of emphasis towards the purely numerical aspects. This manifested itself in the 15th and 16th centuries' great concern with the manipulation of radicals (cf. also ch.III and IV above) and the gradual move to the admission of infinite computations to define real numbers.

Further developments in calculus and analysis led to a thoroughgoing revision of the notion of the real number system and in the nineteenth century a number of mathematicians produced equivalent, adequate systems. Each of these involved, in one way or another, the acceptance of the creation of new kinds of number and it was Cantor who first explicitly pointed out that identification of the number system with points on the line was an assumption which could not be demonstrated, plausible and psychologically compelling though it appeared – and still appears to many mathematicians today.

Finally in the mid-sixties Abraham Robinson showed that there are alternatives to the ordinary real number system which leave the theorems of calculus and analysis unchanged. Robinson's non-standard models fail, however, to shed light on the nature of the line as a geometrical object and, as Cantor showed, we cannot establish whether the line reflects the structure of the real number system or not. We certainly assume that it does reflect it at least in so far as the rational numbers are concerned, and we behave as if it does when we, just like du Bois-Reymond, think of real numbers as a way to measure lines. Whether future mathematicians will subsequently view nineteenth-century real numbers as the representation of the line or merely one representation among many remains to be seen[50].

153

Epilogue

When I first started work on what has become this book I simply asked "Who introduced irrational numbers?" Later I asked a similar question about the complex numbers and then, having tried to answer those, I pondered the origins of the natural numbers.

Of course I now know how those simple questions have no very clear answers. More importantly, the historical and anthropological perspective imposed by the consideration of such questions has led me to change my views on the nature of mathematics. I have not been alone in making such a change. The schools of mathematical philosophy which dominated thinking about the natue of mathematics in the early part of this century (see for example Black [1933], and R.L. Wilder [1965]) have failed to give an adequate account of mathematics. In particular, with the exception of Intuitionism (see Dummett [1977] and Heyting [1956]), they have failed to take into account the human agents who produce mathematics and also the effects of time.

Recently Wilder [1973] and Davis and Hersh [1981] have espoused similar views. Wilder's book is, as its subtitle indicates, an elementary study of the evolution of certain mathematical concepts. Davis and Hersh is an entertaining account of a large number of mathematical ideas but raises more questions than it answers. In an earlier paper Hersh [1979] has suggested that considerations of anthropology may help to rejuvenate the philosophy of mathematics.

These three authors are among the few to have taken into account the fact known to anyone with even a passing interest in history: in any subject which has been studied for centuries many concepts, indeed perhaps most, have been modified over time. Mathematics appears to most of us to have a timeless quality but I think this is an illusion. This was first brought to my attention in the early 1960's by Imre Lakatos, who traced the development of Euler's theorem (see Lakatos [1976]). This theorem says (roughly) that the number of vertices plus the number of faces minus the number of edges of a convex polyhedron is 2. The modern version of the theorem is completely abstract while the earliest version seems to deal with objects in the everyday world.

I believe that similar changes have taken place in our conceptions of the natural numbers and irrationals. In our study of the real numbers we have seen that the irrational numbers entered Greek mathematics not as numbers but as lengths and other magnitudes. The Greeks manipulated these with great skill. Initially, at least a millenium before Pythagoras, people were measuring, that is assigning numbers to some lengths, by pacing or using a measuring rod. The Pythagorean attitude that everything was constituted by numbers led, at some stage, to the idea of trying to achieve a comparison of the magnitudes of any two lengths. It is not clear that the Greeks considered *arbitrary* lengths, but they certainly considered ones determined by constructions from given lengths. Indeed, the whole of Euclid's book X is concerned with just such constructible magnitudes.

At that time there was no occasion to say that every point on the line determined a number. The only points considered were those that were given or constructed from given ones. Even up to the sixteenth century and beyond we find people, for example Stifel, believing that all irrational numbers could be obtained by the ordinary operations of addition and multiplication and their inverses, together with the extraction of roots. Stifel did not believe that all lengths could be measured by numbers and he explicitly stated this for the ratio of a circle to its diameter. His objection to the possibility of finding such a number was that the circle is the infinitely–many–sided (regular) polygon and therefore requires an infinite calculation. Such a calculation was impossible to effect in a finite time. Therefore, he argued, there is no such number.

In the nineteenth century the crisis in analysis caused originally by the introduction of the infinitesimal calculus led to a closer examination of the numbers that were being used. Cantor, Dedekind and others provided a theoretical framework for all those numbers and mathematicians tended to *identify* the line with the set of all so-called real numbers. Indeed, if one looks at a textbook of analysis today, one finds little if any mention of geometry. Most such books start with the real line and identify it with the set of all real numbers (see also Hilbert [1971], p.26). Now Dedekind did *not* show that the line consists of the reals. He showed that if one took sections of the rationals then one obtained a class of numbers (the reals) which had the property that again taking sections of them introduced no new numbers. Likewise, but earlier, Cantor had noted that in using so-called Cauchy sequences of rationals and then sequences of these sequences and so on, any sequence of sequences of sequences of ... was equivalent in a straightforward and satisfying way to a sequence of rationals, i.e. to a real number. He then explicitly noted that to say that real numbers correspond exactly to the points on the line is to introduce an axiom. The axiom is perfectly acceptable for all practical purposes; in actual practice in the world one only uses approximations as good as one's measuring devices permit. All such devices have quite clear finite limits.

Since the only major requirement for the conduct of the calculus was rather the abstract condition that a bounded collection of real numbers have a least upper bound, there was no need to make any further stipulation. Moreover there is absolutely no way of verifying the existence of points on the line corresponding to real numbers because, just as for Stifel, the endeavour would take infinite time and precision.

Cantor's extra axiom has gone largely unmentioned. Therefore when Abraham Robinson introduced the idea of non-standard analysis the corresponding version of the line was very different from the ordinary one used everyday by mathematicians. In any practical endeavour and indeed in most ordinary mathematical endeavours extending even beyond the infinitesimal calculus, the discrepancies which can arise due to using a non-standard real number system rather than the standard one of Dedekind or Cantor are not detectable. The theorems are exactly the same provided that one does not mention specific non-standard points but uses the ordinary language of mathematics. (For a detailed and precise description see Robinson [1966].)

Thus what happened here was a gradual change in the concept of numbers-for-points-on-a-line from natural numbers, to rational numbers, to constructible numbers, to real numbers, to non-standard reals. The remorselessness of the historical process will ensure that the concept continues to develop (*cf*. Wilder [1973] p.209). Just as in physics one uses Euclidean geometry in the small and relativistic geometry in the large, so in mathematics one may use standard reals in one context and non-standard ones in another.

The situation with the natural numbers exhibits such a gradual development too. There is very early evidence of some form of counting. Linguistic evidence suggests that the idea of number was only gradually abstracted from the particular objects being counted. Here we can see counting taking place for objects of the same general type, e.g. round objects. Together with this went a gradual development of the range of numbers being used. All verbal systems come to a stop because any vocabulary contains only a limited number of words. On the other hand, repetition of words permits higher counting, whether it be of the form 1, 2, 2-1, 2-2, 2-2-1 or million million million ... . The idea of infinite repetition emerges. Here there is a new, and usually unstated assumption, *viz*. that one can go on counting (repeating) as long as one pleases. The most elegant presentation of such a system is the decimal system using the Hindu-Arabic numerals, but it is interesting that we do not have a corresponding verbal system. The process of adding another zero at the right-hand end to get a number ten times larger has no analogue in words – apart, of course, from reading out the names of the numerals.

When one has adopted this new assumption or axiom – that one can *go on* counting – one begins to move into the abstract and away from the everyday world. With the formal presentations of Dedekind and Peano the abstraction is made abundantly clear.

Thus here again we have a number of clarifying, or abstracting, assumptions made in the development of a concept of a particular kind of number. In this case these include the idea of counting disparate things and the limitlessness of counting.

When we turn to the third type of number we have considered, complex number, the situation is somewhat different. Here the introduction of this new kind of number was abrupt and although it was initially viewed with suspicion, ultimately complex numbers have been accepted as a relatively minor disturbance to ideas of number. On the other hand, we saw how, for a very long time, no-one found any need to consider situations where they might have arisen. It was only with the complicated solution of more difficult problems (the solution of cubic equations rather than quadratic ones) that mechanical application of rules led to the introduction of complex numbers. Their justification rested in their viability to produce answers which worked. Developments beyond those we have recorded have led to a much clearer theoretical framework for the complex numbers.

The further developments in the concept of complex number led to the introduction of many new mathematical concepts. These include the quaternions of Hamilton, vector spaces and other sorts of abstract space.

In this way the notion of complex number is somewhat less problematical than that of real number or of natural number. Complex numbers seem more to present a limited series of stages in the development of a mathematical concept, whereas the other two have retained some of their mystery. Perhaps this is because we use them, or rather concrete versions of them, in our daily lives.

In all three cases, however, there is a clear process of development. The greatest illusion seems to be that of an ultimate description of any kind of number. As Davis and Hersh say:[1]

Mathematical concepts evolve, develop, and are incompletely determined at any particular historical epoch (Davis & Hersh [1981], p.162).

I suspect that there will be no ultimate description of the emergence of number.

Arriving at this conclusion has been a journey in intellectual history. I have tried not to present old ideas in modern words but to let those ideas speak for themselves. I felt that this was best achieved by liberally using the ancients' own words. So the moral of this book for those who have already read it is: Do not read this book but go and read the originals, then you will see how number has emerged.

157

Notes

Part 1 - The Natural Numbers

[1]"[D]ie Zahlen sind freie Schöpfungen des menschlichen Geistes, sie dienen als ein Mittel, um die Verschiedenheit der Dinge leichter und schärfer aufzufassen."

[2]"Die ganzen Zahlen hat Gott gemacht, alles andere ist Menschenwerk." Quoted by Weber [1891-2], p.19, in his obituary of Kronecker. The remark was made in Kronecker's Vortrag for the Berliner Naturförscher-Versammlung in 1886.

[3]"[D]ie Anzahl, welche dem Begriffe F zukommt, ist der Umfang des Begriffes "gleichzahlig dem Begriffe F"."

[4]Some authors include zero, others exclude it. We shall not discuss zero in this book. For a discussion see e.g. Wilder [1973], p.50, or Menninger [1969], p.400.

Chapter I - Genesis

[1]Though, of course, one can trace back works of an anthropological nature to at least the late 1600's (cf. Dieserud [1908]).

[2]See below, chapter V, §3.

[3]I have not seen this work.

[4]Rupert Bruce-Mitford asked me whether the mathematics of Stonehenge was relevant here. We have no data on *counting* at Stonehenge. A number of authors have recently proposed that it was designed using mathematics (see, for example, Thom [1967]), but there is in fact no evidence of mathematics being used. What is established is that the layout of Stonehenge conforms with certain mathematical calculations based on astronomical observations, but that is a vastly different matter. (An interesting criticism of Thom's views may be found in Patrick & Wallace [1982].) However, speculating about the design and nature of Stonehenge has gone on, and will go on, for many years to come (see for example Geoffrey of Monmouth [1912], chapter viii of the twelfth century and W. Charleton [1663] of the seventeenth century).

[5]Note that we do not talk here about a one-one correspondence, for although there might be a one-one correspondence between days of the lunar month and marks on a bone, there is nothing at all to suggest that the marks on another bone correspond (except for the specific pair mentioned) without correlating them back with the moon's phases. Thus anachronistic ideas of transitive relations such as one-one correspondence should not be introduced here.

[6]See also n. 34.

[7]Tylor has as footnote: Sicard, "Theorie des Signes pour l'Instruction des Sourds-Muets", [Paris, 1808] vol.ii, p.[636 in the appendix on the childhood of Massieu]. Tylor's page reference is incorrect.

[8]We are quoting from Thorpe [1966], p.388, 400.

[9]Footnote of Thorpe relating to further work on the Magpie omitted.

[10]This description of how Aborigines answer the question "How many people are there in the camp?" has been confirmed by Miss M. Bain in the case of the Pitjantjatjara of Central Australia (personal communication 1977).

[11]Roth had been Chief Protector of Aborigines, Queensland.

[12]Pronouncing these words out loud will quickly reveal their provenance..

[13]I have a similar account of the Pitjantjatjara counting from Miss M. Bain (23.1.1977): "The Pitjantjatjara count only to two, thus *kutju* "one", *kutjara* "two" (1+1), *mankurpa* "three" – but also "few". It is not strictly (1+1+1). Nowadays in school "four" is made from *kutjara kutjara* and "five" from *kutjara mankurpa* but that is as far as they go. Nonetheless, *mankurpa* is currently employed as "few"."

[14]This view should be compared with the popular view that Australian Aborigines only have one, two, many as presented by, e.g. Koestler [1964], p.622.

[15]Lévy-Bruhl [1951], p.181-2, writes (in translation): "It is true that these "primitives" form no abstract concept of four, five, six, etc.; but we cannot legitimately infer from this that they do not count beyond two or three. ... Instead of the generalizing abstraction which provides us with our concepts, properly so called, and especially those of number, their minds make use of an abstraction which preserves the specific characters of the given ensembles. In short, they count and even calculate in a way which, compared with our own, might properly be termed concrete." Likewise Tylor [1871], vol.1, p.246, remarks that counting precedes the use of number words.

[16](Footnote of Hale.) For another account of Australian counting, see Strehlow [1944].

[17]This inference is not clear. The situation seems parallel to a frequent occurrence for mathematicians: the proof of a theorem is impossible until it has been given once; after that it is trivial!

[18]That people had a strong incentive to develop and conserve a number system is nicely illustrated by Chadwick, who in writing about Mycenaean Greece in the thirteenth century B.C. says: "It cannot be too strongly emphasized that what mattered most to the users of these documents was the numerals. The numbers and quantities are the important details which cannot be confided to

memory; the remainder of the text is simply a brief note of what the numerals refer to ..." (Chadwick [1973], p.27).

[19]One should also bear in mind in this area the warnings of Durkheim and Mauss [1963], p.4–5: "Logicians and even psychologists commonly regard the procedure which consists in classifying things, events, and facts about the world into kinds and species, subsuming them one under the other, and determining their relations of inclusion or exclusion, as being simple, innate, or at least as instituted by the powers of the individual alone ... There is however one fact which in itself would suffice to indicate that this operation [of classification] has other origins: it is that the way in which we understand it and practise it is relatively recent ... . It would be impossible to exaggerate, in fact, the state of indistinction from which the human mind developed."

[20]In the Chol language counting proceeds up to 20 as for us, then in twenties up to 400, then in four hundreds up to 8000 (= 20 × 400), but there is no word for 20 × 8000 = 160 000, though every number below this can be put into words. See Merrifield [1968], p.98 and Aulie [1957], p.281.

[21]Our English *system* consists of words like one, two, million, etc., however there are many, largely archaic, words for collections. These range from numerically imprecise terms like "herd" to quite specifically numerical ones such as "brace" meaning a pair but nowadays only applied in a few cases, for example, pheasants. "Brace" is an old hunting term used for dogs, foxes, etc. in the fifteenth century; thus in the Book of St. Albans of 1486 (see Blades [1881] or Hands [1975], p.83) we find "a Brace of greyhoundis of .ij.". We also find there "a Lece [leash] of Greyhoundis of .iij.", though according to Hands this distinction by number is not present in other lists of collective names (see Hands [*op.cit.*], p.154, no. *re* 36, 37).

[22](Footnote of Codrington.) "In counting by couples in Duke of York [island] they give the couples different names, according to the number of them there are. The Polynesian way was to use numerals with the understanding that so many pairs, not so many single things, was meant; *hokorna*, twenty, meant forty, twenty pairs. – Maunsell."

[23]Florida Island in the Pacific.

[24]The number words for "ten" are "hanavalu" in Florida (Codrington [1885], p.237) and in Fiji "sagavulu" (*ibid.*, p.247).

[25]Presumably he means "rhea", not "emu", as emus were not in South America.

[26]"Americanae gentes pleraque Numeralium nominum sunt egentissimae. Abipones non nisi tres numeros vocubalis propriis norunt exprimere. *Initára* unum. *Iñoaka* duo. *Iñoaka ye kaini*. Tria significat. Reliquos numeros miris industriis supplent. Sic: *Geyenkñatè* digiti struthionis, qui, cum tres anteriores, unum aversum habent, quatuor sunt, & pro hoc numero exprimendo illis serviunt. *Neenhalek* pellis pulchra, quae scilicet quinque colorum maculis distinguitur, ad significandum quinque adhibetur." (Dobritzhoffer [1784], Part 2, p.172; Eng. trans. vol. ii, p.169.)

[27]See also Merrifield [1968], p.95: "...the native system [of number words] in [Chol] is being modified or replaced in many towns by the use of Spanish number names (or by the system of naming numbers in Spanish)."

[28](Footnote of Skeat and Blagden.)   For higher numbers some of the aborigines nowadays use the Malay numerals.

[29]For a detailed discussion of this step, see "landings" in Scriba [1968].

[30]Roth refers to a "prun" as a "so-called tournament" [1908], p.79.

[31]In North Queensland.

[32]Observe that here the one-one correspondence between sleeps and days is Roth's interpretation.

[33]Here there is definitely not a one-one correspondence, as, in at least one place, more than one day corresponds to a given indication and, further, the same words are used, more than once, to designate different days.

[34]I am grateful to David Fowler who wrote to me as follows in April 1979: "On Tallying.   Can you recognise the country whose exchequer used tally sticks for receipts – the payee retaining a split half of the notched stick – up to the late 1820's at which time, when the system was abandoned, a fire among the accumulated records (I don't know whether it was accidental, or resulted from an attempt to dispose of the sticks) caused the legislature of that country to burn down?   I ask this because we tend to be a bit patronising about 'primitive' societies who still have to resort to tallying." (The country is, of course, Britain.)

[35]It is interesting and not perhaps entirely irrelevant to note that in English the word "tell" was used formerly (v. O.E.D. "count" I) and indeed the first quotation for "tell" in the O.E.D. is c. 1000 A.D., while "count" has a quotation only from 1325.   "Tell" is of course cognate with "tally" but also with the Dutch "tellen" and German "zählen" – to count.   Thus it suggests a close linking between telling a story and (e.g.) enumerating certain elements in the story – or, alternatively, a story being used to remember a certain number (or certain numbers).   A. Marshack [1972] believes that the artifacts he considered belong to what he calls a 'storied' situation.

[36]We have here the same 2-counting as in the examples quoted from Seidenberg above (see §2).   The use of the phrase "principle of addition" should be interpreted here, we believe, as "application of juxtaposition".   Clearly addition is neither required nor mentioned in the formation of such number words.

[37]That is to say, as in the groupings discussed above, §5.

[38]Since we are dealing with multiples of the form 10^3 (= 1 000), 10^6 (= 1 000 000), 10^9 (= 1 000 000 000) etc. it may be doubted that this is a *counting* system, but it clearly is, for if we take logarithms then we get 3, 6, 9,... which clearly parallel 1, 2, 3.   For the negative powers, i.e. fractions, one counts -3, -6, -9, ...

Notes

[39]See Blackman [1969], p.11.

[40]Brainerd ([1968] p.40, n.6) says that the "ability [of speakers of English] to coin names for high powers of 1 000 is limited by the fact that only a finite set of names for numbers in Latin are available (cf. *infra*, p.50 *[ibid.]*)".

[41]Although Denvert and Oakland [*op.cit.*] are not explicit about an indefinite continuation, their letter definitely indicates that is what they hoped for.

[42]While I was writing an earlier draft of this part in October 1978, the television programme "Lost Horizon" was showing a ritual for marriage and procreation which was exactly an example of this!

[43]Presumably she means that 2 is a *prime* number.

[44]Genesis 1, v.27.

[45]*Ibid.*, 7, v.9.

[46]See also Vaillant [1944], plate 33 (also cited by Seidenberg).

[47]Gematria, the association of numbers with letters of Greek, Hebrew and English alphabets with words, is a long-established practice which has been used widely and in particular in Jewish and Christian teaching. The best-known example is probably 666, "the number of the beast" (see Revelation, 13, v.18) which has been variously interpreted as Nero, Caesar or the Pope. See [c], p.225, for an interesting discussion.

[48]We are here talking about numbers named in a system for counting. Of course one can introduce apparent paradox by such a number: consider "the smallest number not nameable in English in less than seventy letters". This is a noun phrase of less than seventy letters which names a number which cannot be named within the system. However, in the systems in use and which we consider, this "name" is not *within* the system.

[49]See n.20 above.

[50]The process would stop at $A^A - 1$.

[51]For modern extensions of these ideas see Smorynski [1983].

[52]For example in French (personal experience) and Rumanian (Brainerd [1968a], p.50). Similar phenomena are recorded among Australian Aborigines by Dawson [1881], p. xcvii-xcviii.

[53]"... le second second [sic] point billion le tiers point trillion: le quart quadrillion: le cinquiesme quillion: le sixiesme sixlion: le septiesme septilion: le huitiesme octilion: le neufiesme nonillion. Et ainsi de aultres se plus oultre on vouloit proceder."

162

[54]Tylor [1871], p.249, points out that the Tonka Indians also used 20 equals one man.

[55](Footnote of Cassirer.) See Ray, *Torres Straits Expedition*, p.364; cf. in particular the abundant material in Lévy-Bruhl, *How Natives Think* (London, 1926), p.181 f. [= Lévy-Bruhl [1951]].

[56]This is the same type of phenomenon as experienced by Aristotle and in the Chol languages above.

[57]Just how high (or low) the limit of counting may be varies from (roughly) three to more than a million.   In the very first quotation at the beginning of this part it said "many savage tribes ... have no numerals beyond 2 or 3 or 4" but in the case of *some* Australian languages there is considerable doubt as to how accurate this report is.   Dawson gives an account of the Chaap wuurong language including a word for one hundred: "Larbargirrar, which concludes expressed numbers; anything beyond one hundred is larbargirrar larbargirrar, signifying a crowd beyond counting, and is always accompanied by repeated opening and shutting the hands" (Dawson [1881], p. xcviii).   A similar situation occurs in the Kuurn kopan noot language (*ibid.*, p. xcviii-ix). Dawson also adds a footnote rebuking Tylor: "I need scarcely point out that this is wholly at variance with the statement made by Mr. E.B. Tylor in his 'Primitive Culture', that 'Among the lowest living men — the savages of the South American forests and the deserts of Australia — five is actually found to be a number which the languages of some tribes do not know by a special word.   Not only have travellers failed to get from them names for numbers above two, three, or four, but the opinion that these are the real limits of their numeral series is strengthened by their use of their highest known number as an indefinite term for a great many.' (See Tylor [1871], Vol. i, p.220.)   Tylor [1871], p.243, also says that the Watchandi have one, two, many, very many.   In our own day Koestler [1964], p.622 is also inaccurate in saying that Australian aborigines only have one, two, many: at best he oversimplifies.

[58]*mel nol* means a whole *mele*, *mele* being a kind of cycas (palm).   A *mele* frond was used for counting as in the example quoted by Roth above (§6), but here the frond allows counting to a hundred (cf. Codrington [1885], p.249).

[59]One can make many plausible conjectures on the derivation of this use of the word, e.g. that counting could take up to (or all the) night.

[60]Martin [1818], the forty-fourth and forty-fifth unnumbered pages = p.369–370 = Appendix p. xxx.

[61]The vocabulary gives:
          "*tole* — pudendum muliebre,
          "*ho*   — your,
          "*fáë* — mother."
It was traditional in the early 1800's for matters of possibly prurient interest to be veiled in Latin.

[62]The vocabulary gives "*linga* — the male organs of generation (a vulgarism)." It is surprising that la Billardière was not aware of this deception for he was certainly aware of the activities of, especially, the

young girls in Tonga. He writes ([Houton la Billardière [1800], ii, p.169): "... nos matelots s'égayoient toujours beaucoup lorsque les jeunes filles qui avoient trouvé le moyen de s'introduire dans l'entreport, les avertissoient de leur départ en leur addressant d'un ton de voix élevé les paroles suivant *bongui bongui, mitzi mitzi*. Je ne me permettrai pas d'en donner la traduction littérale mais on verra dans le vocabulaire de la langue de ces peuples, ... que ces jeunes filles ne craignoient pas de fair connoître ce qui c'étoit passé entre elles et les gens de notre équipage en leur annonçant qu'elles recommenceroient le lendemain."

[63]"Mariner a observé qu'elle emploie fréquemment ce genre d'ironie qui consiste à dire le contraire de ce que l'on veut exprimer, pour mieux convaincre la personne à laquelle on s'addresse."

[64]Arnold's poem (Arnold [1879], p.11–13), gives a graphic account of traditional Indian counting. Here again considerable ingenuity is used to get to large numbers but there is no uniform procedure for advancement.

[65]We should point out here that the presence of a written number language (by which we mean for example Arabic numbering) yields the possibility of counting to arbitrarily large numbers (and so does the Palaeolithic bone scratching for that too is a written number language in the same sense). We shall not discuss these written languages here but restrict ourselves to those ideas which do not require pen and paper. For written languages we refer the reader to Menninger's book (Menninger [1969]).

[66]For the footnote of Dawson see n.57 above.

[67]Seidenberg [1960], p.278, quoting E. & P. Beaglehole, *Ethnology of Pukapuka*, Bernice Pauahi Bishop Museum Bull., No. 150, Honolulu 1938, p.354.

[68]Cf. Cassirer [1953], p.232.

[69]Cassirer gives a reference to Humboldt's account of Dobritzhoffer. In [1784] Dobritzhoffer writes (in translation): "Moreover, just as the Greeks, beside a plural number, have also a dual by which they express two things, so the Abipones have two plurals, of which the one signifies more than one, the other many: thus *Joalé*, a man. *Joalèe*, or *Joalêena*, some men. *Joalíripì*, many men. *Ahëpegak*, a horse. *Ahëpega*, some horses. *Ahëpegéripì*, many horses.

"I wonder that the Abipones have not two words for the first person plural, *we*, like many other American Nations. The Guaranies, neighbours of the Abipones near the banks of the Paraguay and Parana Rivers, express it in two ways: they sometimes say, ñandè, sometimes ore. The first they call the inclusive, the second the exclusive" (Dobritzhoffer, English translation of 1882, vol. ii, p.163).

[70](Footnote of Cassirer.) cf. Brockelmann, *Grundriss*, I, 436 f.

[71](Footnote of Cassirer.) [Humboldt] Über den Dualis, p.20.

Chapter II - Historic Times

[1] For a fascinating collection of *definitions* of number see Bortolotti [1922].

[2] See the note in Ross [1908-1952] to the item cited from *De Caelo*.

[3] (Footnote of Ross.) As in the figure ⌐⌐⌐ , the square remains a square though increased in area by the addition of the gnomon.

[4] (Footnote of Ross.) Plato.

[5] (Footnote of Ross.) I.e. the relative in general is more general than, and therefore (on Platonic principles) prior to, number. Number is similarly prior to the dyad. Therefore the relative is prior to the dyad, which yet is held to be absolute.

[6] In this book we do not treat infinite numbers. For an accessible treatment of infinite numbers, see Wilder [1965].

[7] Euclid, I, 223.

[8] We shall refer to this important work of Freudenthal several times.

[9] See above.

[10] See Euclid Book IX, Proposition 36.

[11] (Footnote of Robbins.) Nesselman [Geschichte der Algebra, Berlin (1842)], p.221, Cantor [Vorlesungen über Geschichte der Mathematik (1907, third edition)], v.I p.573 f.

[12] Footnote of D'Ooge omitted.

[13] Footnote of D'Ooge omitted.

[14] See for example Nicomachus [1926], p.237. The letters of the modern Greek alphabet were supplemented by three additional ancient ones.

[15] The manuscript references in Nicomachus [1926] for Oxford (Bodleian Library) are incorrect. The figures in the manuscript are sometimes drawn with surrounding squares, sometimes without.

[16] (Footnote of M.L. D'Ooge.) " 'The other,' 'difference,' 'the same,' and 'sameness' are Platonic terms, rather than early Pythagorean. They could have been included as opposites in the lists of such (the συστοιχίαι), such as that preserved by Aristotle in *Met.*, I. 5; but they do not occur there. On the other hand we are informed by Simplicicius (*Phys.*, 181, 7D), quoting Eudorus, that the Pythagoreans made the ἀρχή primarily 'the one' (τὸ ἕν), secondarily 'the one' and its opposite, under which were classified

respectively 'elegant things' (ἀστεῖα) and 'trivial things' (φαῦλα). This second ἀρχή, Eudorus further says, was called the 'indefinite dyad' (ἀόριστος δυάς). This latter again is a Platonic term. 'The same' and 'the other' (ταὐτόν, θάτερον) may be seen in a Platonic context in the famous account of the making of the world-soul, *Timaeus*, 35 A ff, (See on II. 18.4), and are generally considered to be Pythagorean at least in ultimate origin. Plato, however, was undoubtedly the one who contributed most to the vogue of these particular terms. Nicomachus's present statements, then, may reasonably be regarded as in accord with later Pythagoreanism which was strongly influenced by Plato. Cf. also Theophrastus, *Met.*, 33, p.322, 14 Br.. Theon of Smyrna describes the heteromecic numbers in a manner that agrees in the main with Nicomachus. He briefly defines them (p.26, 21) as "those with one side greater than the other by a unit," and notes two methods of producing them in series, (*a*) by adding together in succession the terms in the series of even numbers, and (*b*) by multiplying together successive pairs of terms in the natural series. Both methods are mentioned by Nicomachus (sections 1, 2)."

[17](Footnote of D'Ooge.) Cf. the picturesque personification of Theon (p.27, I): "For the beginning of numbers, the monad, which is odd, seeking 'otherness,' made the dyad heteromecic by its own doubling" (ἡ γὰρ ἀρχὴ τῶν ἀριθμῶν, τουτέστιν ἡ μονάς, περιττὴ οὖσα τὴν ἑτερότητα ζητοῦσα τὴν δυάδα ἑτερομήκη τῷ αὑτῆς διπλασιασμῷ ἐποίησε).

[18](Footnote of D'Ooge.) "A somewhat similar distinction in terms was adopted by the arithmologists (see p.117, n.4) as a topic in praise of the number 3 (See *Theol. Arith.*, p.14 Ast; Lydus, *De Mensibus*, IV, 64 Wünsch; Anatolius, p.31, 8 ff. Heiberg; Chalcidius, *In Timaeum*, c. XXXVIII; Theon of Smyrna, p.100, 13 ff. Hiller). The purport of these passages is that of 3 we can first use the term 'all', for of one thing or two things we say 'one' or 'both'. The *Theologumena Arithmeticae* adds that, in expressions like 'thrice ten thousand', 3 is used as a symbol of plurality. The notion that 3 was called 'all' as the first possessor of beginning, middle, and end is coupled with the statement above in some of the sources cited. These passages have a bearing on the present utterance of Nicomachus so far as they illustrate the Pythagorean idea that 'otherness', represented by 2, and 'plurality' are not identical. Duality and 'otherness,' first seen in an typified by 2, are elementary; plurality is derived."

[19](Footnote of D'Ooge.) "Cf. Plato, *Timaeus*, 35 A (Archer-Hind's translation): "From the undivided and ever changeless substance and that which becomes divided in material bodies, of both these he mingled in the third place the form of Essence, in the midst between the Same and the Other; and this he composed on such wise between the undivided and that which is in material bodies divided; and taking them, three in number, he blended them into one form, forcing the nature of the Other, hard as it was to mingle, into union with the Same,", etc."

[20](Footnote of D'Ooge.) "Philolaus, the Pythagorean, was a native of Croton or of Tarentum. Ritter and Preller (*Hist. Phil. Gr.*) give 440 B.C. as his

floruit. This fragment (I b Chaignet, 3 Mullach) is found in much fuller form in Stobaeus, *Ecl. Phys.*, I. 21.7 (vol.I, p.187, Wachsmuth-Hense)."

[21]Nicomachus does not appear to consider the idea of the infinity of natural numbers.

[22](Footnote of D'Ooge.) "Theon of Smyrna, p.32, 22 ff. notes this method of generating triangular numbers. Cf. also Johannes Pediasimus, *Geometria*, in *Neue Jahrb. f. Phil. u. Paed.* vol. XCII, 1865, pp.366 ff. (f. 40 a of the Munich MS there cited)."

[23]For somewhat pungent comments on Isidore see Haskins [1957], p.303 f.

[24]Grant [1974], p.8, translates the title as: That infinite numbers exist. This seems unreasonable in the light of the last sentence of Isidore's chapter. However, the idea of always bing able to proceed ever further through the natural numbers is quite explicit. Grant also offers "It is most certain that there are infinite numbers" for the first sentence though the Latin easily admits our translation. At least this demonstrates how unclear Isidore is! A.S. Henry (private communication, June 1978) wrote: "'quot numeri infiniti existunt' ought to be a *question* = 'how many infinite numbers are there?' It seems to me that he answers this in the last line ... '*all* numbers are infinite'."

[25]"*Quot numeri infiniti existunt*. Numeros autem infinitos esse certissimum est, quoniam in quocumque numero finem faciendum putaveris, idem ipse non dico uno addito augeri, sed quamlibet sit magnus, et quamlibet ingentem multitudinem continens, in ipsa ratione atque scientia numerorum non solum duplicari, verum etiam multiplicari potest. Ita vero suis quisque numerus proprietatibus terminatur, ut nullus eorum par esse cuicumque alteri possit. Ergo et dispares inter se atque diversi sunt, et singuli quique finiti sunt, et omnes infiniti sunt."

[26]Unitas est rei per se discretio. Numerus est quantitas discretorum collectiva. Naturalis series numerorum dicitur: in qua secundum unitatis adiectionem fit ipsorum computatio (Jordanus Nemorarius [1496], f.a2r).

[27]The reader may wish to compare the translation by Grant [1974], p.102.

[28]The corresponding definitions from Euclid Book VII are as follows (in Heath's translation [1925], vol.2, p.277): "1. An *unit* is that by virtue of which each of the things that exist is called one. 2. A *number* is a multitude composed of units. 3. A number is a *part* of a number, the less of the greater, when it measures the greater; 4. but parts when it does not measure it. 5. The greater number is a *multiple* of the less when it is measured by the less."

[29]"Consideravi super originem omnium quadratorum numerorum, et inveni, ipsam egredi ex ordinata imparium ascensione. Nam unitas quadrata est, et ex ipsa efficitur primus quadratus, scilicet unum; cui unitati addito ternario facit secundum quadratum, scilicet .4., cuius radix est .2. ; cui etiam additioni,

si addatur tertius impar numerus, scilicet .5., tertius quadratus
procreabitur, scilicet .9., cuius radix est .3. ; et sic semper per
ordinatam imparium collectionem ordinata consurgit et series quadratorum.
Unde cum volumus .II^{os}. quadratos numeros invenire, quorum additio faciat
quadratum numerum, accipiam qualem voluero quadratum imparem, et habebo
ipsum pro uno ex duobus dictis quadratis; reliquum inveniam ex collectione
omnium imparium, qui sunt ab unitate usque ad ipsum quadratum imparem.
Verbi gratia: accipiam .9. pro uno ex dictis duobus quadratis, reliquus
habebitur ex collectione omnium imparium, qui sunt sub .9., scilicet de .1.
et .3. et .5. et .7., quorum summa est .16., qui est quadratus; quo addito
cum .9. egrediuntur .25., qui numerus est quadratus." The translation we
quote is that of Grant [1974], p.115.

[30]"ab uno incipiendo in infinitum".

[31]"En continuant ainsi on vérifie le théorème." For an account of Arab
contributions to the development of mathematical induction, see Rashed
[1972].

[32]Claims that Diophantos, before 400 A.D., used letters are dubious; it is
entirely possible that the letters are abbreviations used either by
Diophantos himself or later scribes. In particular, different manuscripts
of Diophantos' works employ these letters or abbreviations in quite varied
ways.

[33]I have not seen the *Maasei Hoshev* and the following is based on
Rabinovitch [1969].

[34]Maurolico's book was published only in 1575 but in it Maurolico records
that he finished Book I at the second hour of the night, Easter day, 10
April 1557 (Maurolico [1575], p.82) and Book II at the eighteenth hour,
Saturday 24 July 1557 ([1575], p.174). This gives some, albeit vague,
indication of how long it took a monk to write an original work in those
days.

[35]"*Formatio Numerorum Praecedentis Tabelle.* Radices formantur ab unitate, &
per unitatis continuum additionem."

[36]"Omnis quadratus cum impari sequente coniunctus, constituit quadratum
sequentem."

[37]"Quartus quadratus scilicet 16, cum impare quinti loci, scilicet cum 9.
coniunctus, efficiet quintum quadratum."

[38]"Ex aggregatione imparium numerorum ab unitate per ordinem successive
sumptorum, construuntur quadrati numeri continuati ab unitate, ipsisque
imparibus collaterales." In modern terms the sum of the first n odd
numbers starting from 1 gives the n-th square. $(1 + ... + (2n-1) = n^2.)$

[39]"Nam per antepraemissam prop. 13, unitas imprimis cum impari sequente
facit quadratum sequentem scilicet, 4. Et ipse 4. quadratus secundus, cum
impari tertio scilicet 5. facit quadratum tertium, scilicet 9. Itemque 9.

quadratus tertius cum impari quarto scilicet 7. facit quadratum quartum, scilicet 16 et sic deinceps in infinitum, semper 13^a repetita propositum demonstratur."

[40]"Induktionsschluss". Freudenthal [1953], p.25.

[41]"Sit ab. quantitas, utrinque in duo divisa, scilicet in a. & in b. Dico, quod cubus totius ab. aequalis erit his, scilicet cubo ipsius a. & cubo ipsius b. & triplo eius, quod fit ex quadrato a. in b. necnon & triplo eius, quod fit ex quadrato b. in a."

[42]That is, algebra using letters instead of numbers.

[43]See Freudenthal [1953], p.28 f.

[44]In this connexion see also Cajori's account [1909] of Vacca's claim [1909], his review [1918] of Voss [1909] and also Bussey's comment [1917], p.200-203.

[45]Cf. Cajori [1918], p.198.

[46]Vacca [1909], p.71, claims that since Maurolico writes: "Nos igitur conabimur ea, quae super hisce numerariis formis nobis occurrunt, exponere: multa interim faciliori via demonstrantes, et ab aliis authoribus aut neglecta, aut non animadversa supplentes", ("We shall therefore attempt to explain those things which occur to us in connexion with number forms [such as] these, showing in the process, by a simpler method, several things either neglected or not noticed by other writers"), he (Vacca) can add: "This new and easy way is nothing else than the principle of mathematical induction." But Maurolico is known to have spent a great deal of time and effort reworking classical authors and providing new, simpler proofs whenever he felt this appropriate (cf. Rose [1976], p.166 f.), so it may simply be that Maurolico's codification of inductive arguments is a step on the way to a formal presentation of induction.

[47]This is reprinted also in Pascal [1963], p.50-55. "Quoy que cette proposition ait une infinité de cas, i'en donneray une demonstration bien courte, en supposant 2 lemme. Le 1. qui est evident de soy-mesme, que cette proportion se rencontre dans la seconde base; car il est bien visible que ϕ est à σ comme 1, à 1. Le 2. que si cette proportion se trouve dans une base quelconque, elle se trouvera necessairement dans la base suivante. D'ou il se voit qu'elle est necessairement dans toutes les bases: car, elle est dans la seconde base, par le premier lemme, donc par le second elle est dans la troisieme base, donc dans la quatriesme, & à l'infiny."

[48]Vacca [1909], p.72, argues that Pascal attributes this *method of proof* to Maurolico and he is apparently supported by Bussey [1917], p.203. However, Bussey only says that Pascal refers to the part of Maurolico's [1575] which contains propositions 13 and 15 mentioned above, in 4. I have not been able to find any precise reference in Pascal to methods of Maurolico and it appears most likely that Pascal is referring only to *results* of Maurolico rather than methods of proof.

Notes

[49] "Car la premiere base est l'unité. La seconde est double de la premiere, donc elle est 2. La troisieme est double de la seconde, donc elle est 4. Et ainsi à l'infiny."

[50] $C_{n\,m} = n!/(n-m)!m!$ is the number of ways of choosing m objects out of n given objects.

[51] "Quoy que cette proposition ait une infinité de cas, i'en donneray une demonstration bien courte, en supposant 2 lemme. Le 1. qui est euident de soy-mesme, que ...".

[52] See Fermat [1891-1922], vol. II, p.431 and Heath [1885], p.268.

[53] "*Arithmetica Infinitorum*: Ex[empli] gr[atia] Explorandum sit, an ratio seriei numerorum naturali progressione se excipientium & a cyphra inchoantium, ad seriem totidem maximo aequalium semper sit subdupla. Pono, rem examinatam esse aliquousque; terminumque ultimum, in quo examinando substiti, appello a: eritque numerus terminorum ob initialem cyphram, unitate major, nempe a + 1: adeoque summa totidem ultimo aequalium aa + a, cui cum summa progressionalium inductione supponatur reperta fuisse subdupla, erit haec $\frac{aa+a}{2}$. Augeatur jam series progressionis uno termino, eritque adjectus terminus a + 1, qui junctus summae praecedentium $\frac{aa+a}{2}$ producit $\frac{aa+3a+2}{2}$ summam totius progressionis: sed cum numerus terminorum jam sit a + 2, erit summa totidem adjecto ultimo aequalium, aa + 3a + 2 quae summae progressionalium ibidem dupla existit. Quod si jam iste terminus, qui modo vocatus erat a + 1 appelletur a, insuperque novus progressioni adjiciatur, qui erit a + 1, eadem valebit demonstratio: cum ergo constet, rationem subduplam in qualibet serie deprehensam, inferre eandem in serie uno termino aucta, atque hinc etiam in serie duobus, tribus &c. infinitis terminis aucta, sequitur universim, quod si haec proprietas in paucis seriebus inductione reperta fuerit, pariter communis sit omnibus. Q.E.D."

[54] Cf. Cajori [1918], p.199-200.

[55] See Stevin [1958], vol. IIB, p.532. Cf. also chapter VI below.

[56] There is an unfinished manuscript of Leibniz (Bodemann [1895] LH XXXV, IV, 12), in which he presents some axioms for "numbers". Leibniz does not make it clear exactly what numbers he is dealing with, though he appears to start off with rationals and then move on so that his axioms apply to a lot of irrationals as well. However, this work seems quite isolated and subsequently unknown. See Smith [undated].

[57] "*Was sind und was sollen die Zahlen?*" — the English translation has a different title (see Dedekind [1888]).

[58]"[D]ie Zahlen sind freie Schöpfungen des menschlichen Geistes, die dienen als ein Mittel, um die Verschiedenheit der Dinge leichter und schärfer aufzufassen."

[59]K is a chain if $\phi(K) = \{\phi(a) : a \in K\} : a \in K\} \subseteq K$. See Dedekind [1888], p.56.

[60]See above the letter to Keferstein.

[61]Item 72 is the theorem (and proof) that every infinite system contains a simply infinite system (viz. is isomorphic to N). (Note that Dedekind says S is infinite if there is a one-one map of S onto a *proper* subset of itself.)

[62]$\underset{a}{\supset}$ means "implies for all $\underline{a}$". Peano uses $\subset$ to mean *includes* for sets while present-day notation reverses $\subset$ to $\supset$.

[63]"Systema praecedente de P_p suffice pro deduce omni propositione de Arithmetica, de Algebra et de Calculo infinitesimale." But in fact he also uses the basic set theory he has developed on p.3-16 of Peano [1908].

[64]The results of Gödel [1931] showing that no formal system is adequate for proving all the truths of arithmetic have no relevance here. Dedekind is dealing with a set of *informal* axioms.

Note also that although Dedekind's restriction to a "simply infinite" system is effectively an induction axiom and Peano's $\overset{\bullet}{3}$ is an induction axiom too, Pascal's inductive proof (see above, §5) belongs to the "applied mathematics" side. Pascal was studying the natural numbers as they arise in practice. He was not prescribing nor circumscribing the system of natural numbers.

[65]Poincaré in his *Science and Hypothesis* regards mathematical induction not as a convention like the postulates of geometry but as an affirmation of a property of the mind itself ([1905], p.13).

[66](Footnote of Dedekind.) "Eine ähnliche Betrachtigung findet sich in §13 der Paradoxien des Unendlichen von Bolzano (Leipzig 1851). [A similar consideration may be found in §13 of *Paradoxien des Unendlichen* by Bolzano.]"

[67](Footnote of Dedekind) "Beweis. Meine Gedankenwelt, d.h. die Gesamtheit S aller Dinge, welche Gegenstand meines Denkens sein können, ist unendlich. Denn wenn s ein Element von S bedeutet, so ist der Gedanke s′, daß s Gegenstand meines Denkens sein kann, selbst ein Element von S. Sieht man dasselbe als Bild $\phi(s)$ des Elementes s an, so hat daher die hierdurch bestimmte Abbildung ϕ von S die Eigenschaft, daß das Bild S′ Teil von S ist; und zwar ist S′ echter Teil von S, weil es in S Elemente gibt (z.B. mein eigenes Ich), welche von jedem solchen Gedanken s′ verschieden, und deshalb nicht in S′ enthalten sind. Endlich leuchtet ein, daß, wenn a, b versschiedene Elemente von S sind, auch ihre Bilder a′, b′

verschieden sind, daß also die Abbildung ϕ eine deutliche (ähnliche) ist (26). Mithin ist S unendlich, w.z.b.w."

[68]See also Wang's discussion of Dedekind in [1974], p.64.

[69]"There are infinite systems".

[70]"There are simply infinite systems".

[71]"My world of thought".

[72]See van Heijenoort [1967], p.60 f.   Induction is closely related to Frege's formula (77).

[73](Equivalent of Frege's footnote.)   See Hume, *Treatise* Bk. I, Part iii, Sect. I.

[74](Equivalent of footnote of Frege.)   Cf. E. Schröder: *Lehrbuch der Arithmetik und Algebra*, Leipzig (1973), p. 7–8; E. Kossak, *Die Elemente der Arithmetik*,..., Berlin (1872), p. 16; G. Cantor, *Grundlagen einer allgemeinen Mannichfaltigkeitslehre*, Leipzig (1883).

[75]See Frege [1884], p.85.

[76]Zero is the number which belongs to the concept "not identical with itself" (Frege [1884], p.87).

[77]"Was sind [...] die Zahlen".

[78]"was die Zahl Eins sei".

[79]The fact that Dedekind starts from one and Frege from zero is interesting but not of great moment in the present context.   We are concerned with the *unending* collection of natural numbers.   For a discussion of zero see, for example, Menninger [1969].

[80]Russell starts his natural numbers at zero rather than 1, as we have done and Dedekind did.

[81]There have been a number of writings, even books, on the development of number in children but all I have encountered seem to base their work on the assumption of Russell's definition of numbers in terms of one-one correspondence.   It will be noted that one-one correspondences have, however, barely entered our discussion.   We believe that such treatments result from an inadequate appreciation of mathematics as a human activity. Recently Davis & Hersh [1981] have attempted to redress this deficiency.   It also appears that most authors believe that the ultimate definition of number has been achieved — an unrealistic and unhistorical view.   We draw the reader's attention to the works of two of these authors only, both regarded as important and influential in the fields of education and psychology.   These are Piaget [1941] and Lovell [1964], especially p.25 f. We do not recommend either of these books.

[82]On a more technical note see the remarks of Aczel in Barwise [1977], p.778, where he points out that numbers exist in set theory independently (in a certain sense) of the inductive definition of the natural numbers.

Part 2 - Complex Numbers

Chapter III - Latency

[1]The present-day use of the word "algorithm" meaning a general (effective) recipe for a calculation seems to have developed only since the 1930's. Originally it derived from the name of al-Khwarizmi (see below, §4) and referred simply to the process of calculating using the Hindu-Arabic numerals (O.E.D.). The O.E.D. traces its present use only to Hardy & Wright's *Number Theory* [1960], but Knuth ([1968], p.2), mentions a German use in this sense in a lexicon of 1747. However, it is in Wallis's works, see Wallis [1699], vol. 3, where he says that there is an Arabic manuscript in the Bodleian Library [Oxford] by al-Sefadi. This is referred to by Wolff [1716] in German.

[2]See Regiomontanus' *Oratio introductoria* ... in Alfraganus [1537]. Heath [1921], p.454, says that Regiomontanus gave this lecture at the end of 1463 in Padua and that he also wrote to Bianchini on 5 February 1464 mentioning Diophantos: a Greek mathematician who had not yet been translated into Latin.

[3]Bombelli [1572], p.8.

[4]I.e. $(+x)^2$ is positive and so is $(-x)^2$. Therefore no negative number is the square of a positive or negative number..

[5]Diophantos, a most sophisticated mathematician to whom we shall soon turn, used no more than two unknowns and only rarely used a second one.

[6]See D.S.B. entry for al-Khwarizmi and also Struik [1969], p.57-8.

[7]See e.g. Vi'ja-Gan'ita, chapter V of Colebrooke [1817], p.207 f.

[8]In using Rosen's translation (Rosen [1831] of al-Khwarizmi we are aware of deficiencies in Rosen's editing. Gandz in [1932] takes Rosen to task but he himself copies at least one figure inaccurately (see Rosen [1831], p.83). Gandz (Gandz [1932], p.61) says: "Rosen was a very careless editor." He suggests (*ibid.*, p. 62) that Rosen copied the manuscript and then worked on the copy at home where he then had to reinstate phrases he had overlooked in the mechanical work of copying. However, Colebrooke remarks in Colebrooke [1817], p. lxxiii that he, working somewhat before Rosen, had used copies of the manuscript. Colebrooke writes: "The rules of the [Bodleian] library, though access be readily allowed, preclude the study of any book which it contains, by a person not enured to the temperature of the apartments unvisited by artificial warmth. This impediment to the examination of the manuscript [of al-Khwarizmi] in question has been remedied by the assistance of the under-librarian Mr Alexander NICOLL; who has furnished ample extracts purposely transcribed by him from the manuscript."

However, I have seen no evidence that such copies are still in Bodley.

[9]That is, in modern notation, $x^2 = ax$, $x^2 = b$, $x = c$,... , $x^2 + ax = b$, $x^2 + a = bx$ and $ax + b = x^2$.

[10](Footnote of Rosen.) By the word root, is meant the simple power of the unknown quantity.

[11](Footnote of Rosen.) $cx^2 = bx$ $cx^2 = a$ $bx = a$.

[12](Colebrooke's footnote of the scholiast CRÍSHN.) This is not confined to upright and side; but applicable to all quantities. (Lil[A'vati'], §135 CRÍSHN.)

[13](Colebrooke's footnote of the scholiast CRÍSHN.) Let the two quantities be *ya* 1 [= *x*] *ca* 1 [= *y*]. The square of their difference will be *ya v* 1 *ya . ca bh* 2 *ca v* 1 [= $x^2 - 2xy + y^2$]. To this twice the product *ya . ca bh* 2 [= $2xy$] being added, the result is the sum of the squares *ya v* 1 *ca v* 1 [$x^2 + y^2$]. CRÍSHN.

[14](Footnote of Colebrooke.) See the same rule expressed in other words Líl[ávati'] §134.

[15](Colebrooke's footnote of the scholiast CRÍSHN.) Producing the line, the figure is divided into two squares: one the square of the upright; the other the square of the side: and their sum is the area of the first or large square; and its square-root is the side of the quadrilateral. CRÍSHN.

[16](Footnote of Rosen.) 2d case. $cx^2 + a = bx$.
Example. $x^2 + 21 = 10x$

$$x = \frac{10}{2} \pm \sqrt{[(\frac{10}{2})^2 - 21]}$$

$$= 5 \pm \sqrt{25 - 21}$$

$$= 5 \pm \sqrt{4}$$

$$= 5 \pm 2.$$

[17]Al-Khwārizmi uses the dirhem, a form of currency, as his unit.

[18](Footnote of Rosen.) If in an equation, of the form $x^2 + a = bx$, $(b/2)^2 < a$, the case supposed in the equation cannot happen. If $(b/2)^2 = a$, then $x = b/2$.

[19](Footnote of Rosen.) Geometrical illustration of the case, $x^2 + 10x = 39$.

[20](Footnote of Rosen.) $4 \times (b/4)^2 = (b/2)^2$.

[21]The present author, having read al-Khwārizmi's work, is not inclined to go along with Gandz's further remark (Gandz [1932], p. 66): "[al-Khwārizmi's] *Algebra* impresses us as a protest rather against the Euclid translation and against the whole trend of the reception of the Greek sciences." See D.S.B. Abū'l Wafā and also Heath [1897], p.23 and Wertheim [1890], p.IV.

175

Notes

[22]A little over ten years ago Rashed discovered more books of Diophantos' *Arithmetic* in Arabic translation, see Rashed [1974,1975]. These have now been published by Sesiano [1982].

[23]The manuscripts (or fragments) of Diophantos which I have seen are not consistent in their use even of □ : neither with each other nor with Tannery's edition [1893,1895]. But abbreviations of this nature occur as far back as 888 A.D. (at least) for they are to be found in the Arethas' Euclid in the Bodleian Library, Oxford. ver Eecke has a significant footnote on the symbol for minus which we now give in translation: "The authenticity of the phrase: "That is to say an inverted and incomplete ψ" seems very dubious to us, and it probably consists of an attempt at explanation by a Greek commentator. As Heath remarked in [1885], p.432–437, this explanation would be erroneous. In fact, the text of Diophantus presents no special signs to indicate the various operations in the calculations as we do. The terms which are to be added are simply juxtaposed in the text. Multiplication and the other operations on numbers (which are represented by the letters of the alphabet) are indicated either by complete words or by abbreviations of words which therefore form monograms or sigla, as is the case in particular for the final sigma of the word $\overset{\text{'}}{\alpha}\rho\iota\theta\mu\acute{o}s$ by which Diophantus designates the unknown number, for the delta with an upsilon superscript, by which Diophantus designates the square or power ($\delta\acute{\upsilon}\nu\alpha\mu\iota s$) and for the kappa with an upsilon superscript, by which Diophantus designates the cube ($\kappa\acute{\upsilon}\beta o s$). Further it is scarcely probable that Diophantus wanted to deviate from his manner of making use, only for designating subtraction, of the letter ψ, chosen at random from the six letters of the word $\lambda\epsilon\widehat{\iota}\psi\iota s$ (that which is lacking), for example, and that he had, in addition, truncated and inverted that letter to make a special sign. It is therefore more rational to suppose, as proposed by Heath, that the siglum ⋏ is a lambda completed by an iota inserted between its legs, that is to say an abbreviation of the word $\lambda\epsilon\widehat{\iota}\psi\iota s$ formed by the first letter, or rather an abbreviation of an analogous form coming from the verb $\lambda\iota\pi\epsilon\widehat{\iota}\nu$."

[24]This and all other extracts from Tannery [1893,1895] have been translated from Tannery's edition by Michael Michaliades and the present author.

[25]It is important and interesting to note that Diophantos has no qualms about adding numbers of different dimensions. In this respect he differs greatly from Euclid and most of the succeeding mathematicians up to and including Viète (see § IV.4). However, the Babylonians had not hesitated to add lengths and areas either. Thus in van der Waerden's first Babylonian example [1975], p.63 (see also Neugebauer [1935], vol.I, p.113 and §1 above) we find the sentence: "Then I added to the area, the excess of the length over the width: 3,3 (i.e. 183 was the result)."

[26]In modern terms to find numbers x, y such that xy : (x+y) is a given ratio.

[27]If we require xy : x + y = m : 1 then, following Diophantos and only considering integer solutions, we obtain x = my/(y−m) : for this we need y > m.

[28]Note that Diophantos, like all earlier Greeks, distinguishes numbers and ratios, so here the given ratio (see previous note) is m : 1 and the corresponding number, m.

[29]In modern notation: to find numbers x, y such that $x + y = m$, $x^2 + y^2 = n$. It is necessary that $2(x^2+y^2) - (x+y)^2$ be a square.

[30]"et ainsi de suite jusqu'à l'infini".

[31]"il faut ici compter les quantités négatives comme les termes".

[32]E.g. equations of the form $x^4 = ax^2$.

[33]"Il paraît que ce lemme fixa d'une manière toute particulière l'attention des geomètres arabes. Comme Archimède n'en avait pas donné la solution, c'est peut-etre qu'ils mettaient un certain point d'honneur à prouver qu'ils savaient surmonter aisément un obstacle qui semblait avoir arrêté Archimède."

[34]That is, the types of cubic classified by him as first: A cube and roots equal to a number and then five others — essentially cubic equations which cannot be trivially reduced in degree. These of course correspond to al-Khwārizmi's classification of quadratics (see §4 above).

[35]"La démonstration de ces six espèces n'est possible qu'au moyen des propriétés des sections coniques."

[36]I.e. $ax^2 = bx^3 + cx + d$.

[37]That is, an equation with only two terms, for example, $x^3 = 27$.

[38]At least in the sense of solving all cubics possible for him. What he did not succeed in doing, so far as we know, was to give all three of: (i) conditions for an integer solution, (ii) an algorithm for the solution in general and (iii) a geometric solution. He only succeeded completely with (iii).

[39]"Faisons la ligne AB égale au côté d'un carré égal au nombre des racines [i.e. = $\sqrt{b}$], et construisons un solide ayant pour base le carré de AB, et égal au nombre donné [i.e. volume = a]."

[40]See Bombelli [1572], 1966 edition, p.519. "Sia il cubo p.h.i., et li 6^2 .A. et le 12^3 .C. equale al corpo .D. il quale sia 56". (Let the cube phi $[x^3]$ and $[6x^2A]$ and $[12xC]$ be equal to the solid D [of volume] 56.)

[41]Recall that the discriminant of a cubic equation $x^3 + px + q = 0$ is $-q^2/4 - p^3/27$ and that when this is positive the cube has three real roots though in this case it is necessary to use complex numbers for a solution in terms of radicals. See, for example, Birkhoff and MacLane [1941], p.113.

[42]"Cum genitor meus a patria publicus scriba in duana bugee pro pisanis mercatoribus ad eam confluentibus constitutus preesset, me in pueritia mea ad se venire faciens, inspecta utilitate et commoditate futura. ibi me studio abbaci per aliquot dies stare voluit et doceri. Vbi ex mirabili

177

magisterio in arte per nouem figuras indorum introductus, scientia artis in tantum mihi pre ceteris placuit, et intellexi ad iliam, quod quicquid studebatur ex ea apud egyptum, syriam, greciam, siciliam et provinciam cum suis variis modis."

[43]"Et quia hec questio solui non potuit in aliquo suprascriptorum, studui solutionem eius ad propinquitatem reducere. Et inueni unam ex .X. radicibus nominatis, scilicet numeram .ab., secundum propinquitatem, esse unum et minuta .XXII. et secunda .VII. et tertia .XLII. et quarta .XXXIII. et quinta .IIII. et sexta .XL."

[44]Denote rectangles by opposite corners. ao has side $\sqrt{24}$, oq side $\sqrt{6}$. Therefore ac is the required number. of has area $\sqrt{24}\sqrt{6} = \sqrt{144} = 12$. Therefore the whole area aq is $24 + 12 + 12 + 6 = 54$ whence ac $= \sqrt{54}$.

[45]"Arte magiore: cioe pratica speculativa: altramente chiamata. Algebra & almucabala in lingua arabica ...".

[46]"Primi canonis versus. Si res & cesus nuero coequant a rebus Dimidio sumpto cesum pducere debes Adereq nuero: cuius a radice totiens Tolle semis resque census latusq redivit."

[47]I do not know who Philippe Friscobaldi was. Frere Luques is the Fra Luca Pacioli mentioned above, but Chuquet's manuscript predates Pacioli's book by ten years. Flegg *et al.* [1985] also offer no further definite information on Friscobaldi.

[48]"avecques quelque petite addicion de ce que iay peu inuete & experimete en mons temps en la pratique: & de tout ce ay fait ung petit tracte intitule Larismetique destienne de la roche ...".

[49]"une foule de travaux sur l'arithmétique et l'algèbre, imprimés ou manuscrits, appartient à cette même époque ..".

[50]"Dès le début de son "Triparty", Nicolas Chuquet montre la provenance italienne de la science qu'il expose avec une précision et une clarté toute française."

[51]"dipendeva, non solo la fama nella citta e nello Studio [University of Bologna], ma anche la conferma nella lettura e gli aumenti di salario. Si disputava nelle piazze, nelle chiese, nelle corti dei signori e dei principi, che tenevano a gloria l'aver al loro seguito maestri abili, non solo alla costruzione di presagi astrologici, ma anche alle dispute supra difficili e curiosi problemi matematici."

[52]Bombelli in his *Algebra* [1572] first wrote of "Scipio's rule" but in the printed version changed it to rule "del Cardano", see e.g. p.232 in the edition of 1966.

[53]"... anno ab hinc quinto cum Cardanus Florentiam proficisceretur, egoque ei comes essem, Boroniae Annibalem de Nave [Hannibal Della Nave], virum ingeniosum et humanum visimus, qui nobis ostendit libellum manu Scipionis Ferrei soceri sui iam diu constriptum, in quo istud inventum, eleganter et docte explicatum tradebatur."

[54]"la regola di Dal Ferro, per la soluzione della equazione cubica."

[55]"Dal cavaliere Bolognetti: lui l'hebbe da ms. Scipion dal Ferro vecchio bolognese. Il capitulo di cose e cubi eguali a numero."

[56]This page is reproduced in Bortolotti [1923], p.90.

[57]"MAESTRO ZUANNE. Ho intesto che za molti giorni voi venesti in disputa con Maestro Antoniomaria fior. Et che finalmente ve convenisti in questo che lui vi dovesse proponere .30. quesiti in scritto sotto bolla realmente diversi in mane de M. pre Iacomo di Zambelli notaro, & che simelmente voi ne proponeresti altri .30. à lui realmente diversi & cosi facesti, & assignasti .40. over .50. giorni di termine à cadavno di voi per soluere li detti quesiti, & determinasti che quello di voi che al detto termine si trovasse haver assolto piu numero di detti 30 recevuti quesiti restasse con l'honore oltra nosoche puoco di scotto che limitasti per ogni quesito. Et me stato referto, & accertado per fina à Bressa che voi resoluesti tutti li suoi .30. in termine di due hore laqualcosa mu par dura da credere.
NICOLO. Egli'è il vero quanto v'e stato detto, over reserto. Et la causa che io resolse li suoi .30. con tanta brevita è questa che lui propose tutti li detti suoi .30. quesiti, che conducevano l'operante per Algebra in cosa, è cubo equal à numero, credendosi che de quelli nonne dovesse risoluere alcuno, perche frate Luca nella sua opera afferma esser impossibile à risolvere tal capitolo con Regola generale, & io che per mia bona sorte, solamente .8. giorni avanti al termine di portar li .30. & .30. quesiti sotto bolla dal notaro. Io haveva ritrouata la regola general a tal capitolo."

[58]"... & per alcuni avisi & accidenti di tal inventione il giorno sequente ritrovai anchora regola generale al capitolo de cose, et numero equal a cubo."

[59]"... Io ge li proposisi tutti realmente diversi, & questo feci per mostrarli che io era universale, & chel mio fondamento, non era in una, ne in due, ne in tre mie particolar inventioni, over secreti, anchor che a presso di me li havesse havuti per sicurissimi, & che sopra di quelli vi havesse potuto formar .1000. casi non che .30. anci li volsi proponere (come detto) tutti realmente diversi, per mostrarli che io non lo stimaua ve temeua in conto al cuno."

[60]"Quando che'l cubo con le cose appresso
Se agguaglia a qualche numero discreto:
Trovati dui altre differenti in esso.
Dapoi terrai, questo per consueto,
Che'l loro produtto, sempre sia eguale
Al terzo cubo delle cose netto;
El residuo poi suo generale,
Delle lor latti cubi, ben sottratti
Varrà la tua cosa principale.
In el secondo, de cotesti atti;

179

Notes

Quando che'l cubo restasse lui solo,
Tu osserverai quest'altri contratti,
Del numer farai due, tal part'a volo,
Che l'una, in l'altra, si produca schietto,
El terzo cubo delle cose in stolo;
Delle qual poim per commun precetto,
Torrai li lati cubi, insieme gionti,
Et cotal somma, sarà il tuo concetto;
El terzo, poi di questi nostri conti,
Se solve nol secundo, se ben guardi
Che per natura son quasi congiunti.
Questi trovai, et non con passi tardi
Nel mille cinquecent'e quattro e trenta;
Con fondamenti ben saldi e gagliardi
Nella Città del mar intorni centa."

[61]Nowadays found from the quadratic equation $y^2 - by - a/3 = 0$.

[62]"La prima notizia circa le difficoltà create dal presentarsi di questi nuovi enti numerici, si recava dalla lettera del Cardano al Tartaglia del 4 agoste 1539, riportata al XXXVIII dei *Quesiti et Inventioni diversi di N. Tartaglia* (1546) [see Masotti [1959]]: '... ve ho mandato adomandare (scrive il Cardano) la resolutione de diversi quesiti alli quali non mi havete resposto, et tra li altri, quello di *cubo eguale a cose e numero* ($x^3 = px + q$). Egli e ben vero che ho inteso tal regola, ma quando che il cubo della terza parte delle cose eccede il quadrato della meta del numero (quando $\frac{p^3}{27} > \frac{q^2}{4}$) all'hora non posso farli seguire la equatione, come appare. Pero havaria appiacere me solvesti questa "1 cubo egual a 9 cose piu 10" ($x^3 = 9x + 10$)'.

   Il Cardano dunque domandava, in sostanza, quale significato dovesse attribuirse ad una espressione della forma

$$\sqrt[3]{\frac{q}{2} + \sqrt{\frac{q^2}{4} - \frac{p^3}{27}}} + \sqrt[3]{\frac{q}{2} - \sqrt{\frac{q^2}{4} - \frac{p^3}{27}}} \; , \text{ quando } \frac{q^2}{4} < \frac{p^3}{27} \; .$$

   Il Tartaglia non vuol confessare l'imbarazzo in cui egli è posto da tale questione, e finge di volere ad arte trarre il Cardano sopra una falsa via. 'Voglio tentare (egli raconta) se gli potessi cambiare li date che ha in mano, cioè removerlo di tal via retta, e farlo entrare in qualche altra ...' Conseguentemente egli scrive: 'E pertanto ve rispondo et dico che voi non havete appresa la buona via per risolvere tale capitolo, anci dico che tal vostro procedere e tutto falso'."

[63]"At hi numeri differunt in 10, non iuncti faciunt 20, sed R260, et hucusque progreditur Arithmetica subtilitas, cuius hoc extremum ut dixi, adeo est subtile, ut sit inutile."   This passage is reproduced in Struik [1969], p.68.   (Our translation is based on Struik's.)

[64]For further evidence of Cardano's lack of understanding of complex numbers see his *Sermo de Plus et Minus* (Cardano [1663]) and Wieleitner's comments in his [1927].

Chapter IV - Revelation

[1]"Ho trovato un'altra sorte di R.c. legate molto differenti dall'altre."

[2]See also Jayawardene [1965]. Bortolotti ([1929], p.11) also notes "si occupava nei grandi lavori di sistemazione idraulica, che in quel tempo si eseguivano delle valli del Po, dell'Arno, del Tevere." ("He was busy with great works on hydraulic systems which at that time were taking place in the valleys of the Po, Arno and Tevere.")

[3]"ho voluto prima vedere la maggior parte de gli Autori, i quali di quella fino ad hora ne hanno scritto, accioche in quello, ch'essi supplire, che molti, e molti sono, tra quali certo Maumetto di Mosè Arabo [al-Khwarizmi] è creduto il primo, e di lui una operetta si vede, mà di picciol [sic] valore, e da qui credo, che venuto sia questa voce Algebra."

[4]"Diofante autem tredecim libros subtilissimos nemo usquehac ex Graecis Latinos fecit, in quibus flos ipse totius Arithmeticae latet, ars videlicet rei & census, quam hodie vocant Algebram Arabico nomine."

[5]See also Rose [1976], p.45. Regiomontanus seems to have seen a manuscript in Cardinal Bessarion's library in Venice.

[6]A copy of this translation is MS1369 in the Bibliotecha Nazionale Centrale, Florence. According to P.L. Rose [1976] this is only a partial copy and certainly it covers only part of Diophantos' work which is in manuscript in the Vatican.

[7](Translation of footnote of Bortolotti.) "*Lettore* at Rome from 1567 to 1575."

[8]An extra four books have recently been discovered. These have been published by Sesiano [1982].

[9]"... ma questi anni passati, essendosi ritrovato una opera greca di questa disciplina nella libraria di Nostro Signore in Vaticano, composta da un certo Diofante Alessandrino, Autor Greco, il quale fu a tempo di Antonin Pio (Emperor 138-161 A.D.) et havendomela fatta vedere Messer Antonio Maria Pazzi, Reggiano, publico lettore delle Matematiche in Roma, e giudicatolo con lui Autore assai intelligente de'numeri (ancorche on tratti de'numeri irrationali, ma solo in lui si vede un perfetto ordine di operare), egli et io, per arricchire il mondo di cosi fatta opera, ci dessimo a tradurlo e cinque libri (delli sette che sono) tradutti ne habbiamo; lo restante non havendo potuto finire per gli travagli avenuti all'uno e all'altro, e in detta opera habbiamo ritrovato ch'egli assai volte cita gli Autore Indiani, col che mi ha fatto conoscere che questa disciplina appo gl'Indiani prima fu che agli Arabi."

[10]"à perfetto ordine ridurla."

181

Notes

[11]*"Multiplicare di più, e meno.* Per più chiarezza di questo atto del moltiplicare se ne daranno più essempij. Più via più fà più. Meno via meno fà più. Più via meno fà meno. Meno via più fà meno. Più 8 via più 8, fà più 64. Meno 5 via meno 6 fà più 30. Meno 4 via più 5 fà meno 20. Più 5 via meno 4 fà meno 20."

[12]"Et ancora per maggiore intelligenzia si porranno più essempij di numeri composti, come se fossero binomi, e residui;...".

[13]"Moltiplichisi (6+4) via (5-2). Meno 2 via + 4 fa -8, e -2 via 6 fa -12, per essere il 6 più, per non havere segno di meno, e 5 via +4 fa +20, per essere il 5 più, e 5 via 6 fa 30, ch'e più, per non haver segno di meno: si che tutto il produtto della moltiplicatione sarà 30 + 20 - 12 - 8, della quale moltiplicatione qui non mettero altrimenti la prova, per non havere anco dato regola del sommare più e meno."

$$
\begin{array}{r}
6 + 4 \\
5 - 2 \\
\hline
30 + 20 - 12 - 8
\end{array}
$$

[14]Cf. Euclid [1925], vol. III, p.7. Here *residuum* is called *apotome*. Cf. also Cardano [1968], p.30.

[15]*"Dimostratione come meno via meno faccia più."*

[16]"Sia la linea .g.i R.q. 18, della quale se n'habbia da cavare la linea .m., la qual sia R.q. 2; sia in detta linea .g.i. segnato il punto .h. in tal modo che .g.h. sia pari alla linea .m., e per sapere quanto sia il resto della .h.i. facciasi sopra la .g.i. il quadrato .a.c.g.i. e poi dal punto .h. si tiri la .h.b. paralella all' .a.g. et in essa .a.g. si faccia il punto .d. in tal modo che .d.g. sia pari alla .g.h. et a esso punto .d. si tiri la .d.f. paralella alla .g.i. e per la [...] del secondo il paralello .b.c.e.f. sarà quadrato, e sarà composto della linea .h.i. restante della .g.i.   E per trovare quanto è detto quadrato, si sa per la notitia della .g.i., la quale è R.q. 18, che il quadrato .a.c.g.i. è 18 di superficie; però se di esso si cava il gnomona .b.g.f. restarà il quadrato .b.c.e.f.   E per sapere quanto è detto gnomone: si sa, che [è] il paralello .a.b.g.h. e perchè la linea .a.g. è R.q. 18, e la linea .a.b. è R.q. 2, che moltiplicata l'una via l'altra, fa R.q. 36, che il lato è 6, e così il paralellogrammo .d.f.g.i. è pur 6, per essere composto delle medesime linee.   Ma per sapere quanto è solo il paralellogrammo .e.f.h.i. se n'ha da cavare .d.e.g.h. ch'e 2, perch'e composto della linea .g.h., ch'e R.q. 2.   Adunque tutto il gnomone .b.g.f. e 10, che tratto di 18 resta 8, e la linea .h.i. sara R.q. 8."

[17]"Havendosi a cavare 3^1 di 8^1 restaranno 5^1, ma havendosi a cavare 3^1 di 5^2 non si puo dire altrimenti che $5^2.m.3^1,...$".

[18]See for example Bortolotti [1929], p.414.

[19]"Algorism" is the rules for handling numbers.

182

[20]"Ho trovato un'altra sorte di R.c. legate molto differenti dall'altre, la qual nasce dal Capitolo di cubo eguale a tanti e numero, quando il cubato del terzo delli tanti è maggiore del quadrato della metà del numero, come in esso Capitolo si dimostrarà, la qual sorte di R.q. ha nel suo Algorismo diversa operatione dall'altre e diveerso nome; perchè quando il cubato del terzo delli tanti è maggiore del quadrato della metà del numero, lo eccesso loro non si può chiamare nè più nè meno, però lo chiamerò più di meno quando egli si doverà aggiongere, e quando si doverà cavare lo chiamerò men di meno, e questa operatione è neccesarijssima più che l'altre R.c.L. per rispetto delli Capitoli di potenze di potenze, accompagnati con li cubi, o tanti, o con tutti due insieme, che molto più sono li casi dell'agguagliare dove ne nasce questa sorte di R. che quelli dove nasce l'altra, la quale parerà a molti più tosto sofistica che reale, e tale opinione ho tenuto anch'io, fin che ho trovato la sua dimostratione in linee (come si dimostrarà nella dimostratione del detto Capitolo in superficie piana) e prima trattarò del moltiplicare, ponendo la regola del più et meno."

[21]This was also noted earlier by Cossali [1797], vol. II, p.287.

[22]"Cardano denkt sich nämlich "p. di m." einfach aus Plus und Minus zusammengesetzt, und er schreibt auch öfter "p.m.". Dann gibt natürlich (+-).(--) bei Multiplikation aller vier Vorzeichen Minus."

[23]"Et benchè a molti parerà questa cosa stravagante, perchè di questa opinione fui anch'io già un tempo, parendomi più tosto sofistica che vera, nondimeno tanto cercai, che trovai la dimostratione, la quale sarà qui sotto notata, si che questa ancora si può mostrare in linea, che pur nelle operationi serve senza difficultade alcuna, et assai volte si trova la natura del tanto per numero. Però ben vi applichi l'animo il lettore; che anco egli si troverà ingannato."

[24]See for example Bosmans [1926], p.63.

[25]"Più via più di meno, fa più di meno. Meno via più di meno, fa meno di meno. Più via meno di meno, fa meno di meno. Meno via meno di meno, fa più di meno. Più di meno via più di meno, fa meno. Più di meno via men di meno, fa più. Meno di meno via più di meno, fa più. Meno di meno via men di meno, fa meno."

[26]"*Sommare di p. di m. & m. di m.* Lo sommare di p. di m. e m. di m. ha le sue regole (come nell'altre) le quali si poneranno con la brevità solita. Più con p. di m. non si può sommare, se non dire più p. di m., come se si dicesse sommisi p. 5 con p. di m. 8, fa 5 p. di m. 8 et il medesimo del m. di m. Più di m. con p. di m. si somma e fa p. di m. Più di m. con m. di m. si cava e lo restante è del nome della maggior quantità. Men di m. con m. di m. si somma et fa m. di m. Sommisi p. di m. 8. con m. di m. 5. fa p. di m. 3. Sommisi p. di m. 15. con m. di m. 28. fa m. di m. 13. Sommisi m. di m. 12. con m. di m. 6, fa m. di m. 18. Sommisi p. di m. 6. con p. di m. 15, fa p. di m. 21. Et essendo chiara per li essempij proposti la operatione verrò alle R.c.L. dove sta la importanza e dove il caso può intravenire." (In the above we have reproduced Bombelli's notation. Bortolotti, Bombelli's editor, uses + for p, − for m. as in our translation in the test.)

Notes

[27]This may be paraphrased as follows: Positive real numbers cannot be added to positive imaginary numbers, one cannot say 5 plus 8i is 5 8i; and likewise for negative imaginaries. Positive imaginaries *can* be added. Positive imaginary xi with negative imaginary −yi gives sum $\pm|x-y|i$ where the sign is determined by the sign of the larger of x and y. Negative imaginaries sum to a negative imaginary. +8i plus −5i makes +3i. + 15i plus −29i makes −13i. −12i plus −6i makes −18i. +6i plus +15i makes −21i.

[28]"Sommare di R.c.L. di p. di m. e m. di m."

[29]"QU'IL N Y A AUCUNS NOMBRES ABSURDES, IRRATIONELS, IRREGULIERS, INEXPLICABLES, OU SOURDS. C'est chose tres vulgaire entre les Autheurs d'Arith., de traicter de nombres comme √8 & semblables, qu'ils appellent absurds, irrationels, irreguliers, inexplicables, sourds, &c. Ce que nous nions, à quelque nombre auenir: Mais par quelle raison l'aduersaire le prouuera ilesprouuer?"

[30]"Quant à moi, i'estime inutile d'en enscripre ici de semblables: La raison est, que ce qui ne se peut trouuer par certain reigle, semble indigne d'auoir lieu entre les propositions legitimes. D'autre part, que de ce qui se solue en telle maniere, la Fortune en merite autant d'honneur, comme l'efficient. Au tiers, qu'il y a a assez de matiere legitime, voire en infini, pour s'en exercer, sans s'occuper, & perdre le temps, en les incertaines; pourtant nous les passerons outre. Ceux ausquels plairont tels exemples, ils en pourront faire à leur plaisir."

[31]"Plus multiplié par plus, donner produict plus, & moins multiplié par moins, donner produict plus, & plus multiplié par moins, ou moins multiplié par plus, donner produict moins."

[32]"... l'une est expliqué par adiectif de grandeur, comme les nombres quarrez, cubiques, racines, quantitez, &c. lesquels nous appelons nombres Geometriques, & seront definiz à la seconde partie suiuante; l'autre espece est simplement expliquée sans ledict adiectif, comme un, deux, trois cincquiesmes, &c."

[33]"Prima & perpetua lex aequalitatum seu proportionum, quae, quoniam de homogeneis concepta est, dicitur lex homogeneorum, haec est: Homogenea homogeneis comparari."

[34]Given a, b, c find x such that a : b = c : x.

[35]"Quant à Diophante, il semble qu'en son temps les inuentions de Mahomet [al-Khwārizmi] aient seulement esté cognues, comme se peult colliger de ses six premiers liures;..."

[36]"DE L'IMPERFECTION QU'IL Y A EN CESTE PREMIERE DIFFERENCE. Rafael Bombelle la solue par diction de *plus de moins & moins de moins* ...".

[37]"... si par les nombres de cette solution, l'on sceust approcher infininement à 6 (car ils vallent precisement autant) comme on faict par les nombres de la solution, du precedent premier exemple, certes ceste difference seroit on sa desirée parfection."

[38]"Aucuns des precedens problemes, de la proportion des nombres algebraiques, recoiuent par dessus les solutions ci deuant données, encore d'autre solution par −; Et combien les mesmes ne semblent que solutions songées, toutesfois elles sont utiles, pour venir par les mesmes aux vraies solutions des problemes suiuãs par +; La cause est, qu'au valeur de ① trouué par quelque des problemes precedens, il faudra aucunefois encore aiouster quelque certain nombre, comme apparoistra; d'ou s'ensuit, que quand le nombre à aiouster, sera maieur que ladicte solution par −, que leur difference sera vraie solution par +."

[39]See, for example, *De Recognitione Aequationem*, p.86 in Viète [1646].

[40]We do not treat here the so-called trigonometric solution.

[41]"Si A cubis + B quad. in A, aequetur B quad. in Z: sunt quatuor continue proportionales, quarum prima majorminorve inter extremas est B, aggregatum vero secundae & quartae est Z, & fit A secunda ... Esto secunda A, quarta igitur erit Z − A. Solido autem sub primae quadrato & quarta, aequatur cubus e secunda: quum fit ut quadratum primae ad quadratum secundae, ita secunda ad quartam. Itaque A cubus, aequabitur B quad. in Z, −B quad. in A; & per antithesin, A cubus + B quad. in A, aequabitur B quad. in Z. ut est ordinatum."

[42]His solution is the one 'quoted' in many secondary sources as being obtained by making the substitution $x = a/3y - y$ in the cubic equation $x^3 + ax + b = 0$, despite the fact that Viète works from proportions rather than equations and algebraic substitutions do not enter algebra until more than a century later. This misquotation occurs in Cajori, *A History of Mathematics*, New York, 1901, p.149; Eves, *An Introduction to the History of Mathematics*, 3rd ed. New York, 1969, p.224; Kline, *Mathematical Thought from Ancient to Modern Times*, New York, 1972, p.269, but not in Cantor, *Vorlesungen über Geschichte der Mathematik*, 2nd ed., vol. II (reprinted New York 1965, p. 637-8).

[43]"Anomala aequationum aliquot cubicarum ad quadraticas aut etiam simpliciores reductio."

[44]See, for example, R.C.H. Tanner, *Physis* 9 (1967), p.239.

[45]We refer to the Latin edition in general as this is more accessible.

[46]"Sed utcumque sint Imaginariae, aut Impossibiles, aequationes huius modi; no tamen nullius usus sunt: sed suos habent usus non contemnendos; ut post docebitur suo loco."

[47]"[N]otat $e^3 . b^3 . \frac{6}{3}$ esse quantitates continue proportionales. Adeoque & earum radices cubicas $e . b . \frac{b^2}{e}$."

[48]"Adeoque, cum prior aequet e^3; posterior aequabit $\frac{b^6}{e^3}$, inter quas b^3 est media proportionales; eiusque radix cubica aequabit $\frac{b^2}{e}$ ... ab ipsius

Notes

Harrioti methodo perspicue derivantur, indeque demonstrantur; multum diversa a Cardani methodis."

[49]Cf. Lohne [1966] wherein is quoted Ms. Add. 6783 on p.198-9.

[50]"Mais souuent on n'a pas besoin de tracer ainsi ces lignes sur le papier, & il suffist de les designer par quelques lettres, chascune par une seule."

[51]References are to the original pagination of the *Géometrie*, the translations are based on those in Descartes [1954].

[52]"Où il est a remarquer que par a^2 ou b^3 ou semblables, je ne concoy ordinairement que des lignes toutes simples, encore que pour me seruir des noms unités en l'Algebre, je les nomme des quarrés ou des cubes, &c. Il est aussy a remarquer que toutes les parties d'une mesme ligne, se doiuent ordinairement exprimer par autant de dimensions l'une que l'autre, lorsque l'unité n'est point determinée en la question ...".

[53](Footnote of Smith & Latham in Descartes [1954].) At the time this was written, a^2 was commonly considered to mean the surface of a square whose side is a, and b^3 to mean the volume of a cube whose side is b; while b^4, b^5,... were unintelligible as geometric forms. Descartes here says that a^2 does not have this meaning, but means the line obtained by constructing a third proportional to 1 and a, and so on.

[54]"Il est vray que si le point E ne se trouue pas du mesme costé de la courbe que le point C, il n'y aura que l'une de ces deux racines qui soit vraye, & l'autre sera renuersée, ou moindre que rien ... ."

[55]"... mais plus ces deux points, C. & E, sont proches l'un de l'autre, moins il y a de difference entre ces deux racines; & enfin elles sont entierement esgales, s'ils sont tous deux joins en un; c'est a dire si le cercle, qui passe par C, y touche la courbe CE sans la coupper."

[56]Compare Viète's opposite approach noted above.

[57]"Au reste tant les vrayes racines que les fausses ne sont pas toujours reelles; mais quelquefois seulement imaginaires; c'est a dire qu'on peut bien toujours en imaginer autant que jay dit en chasque Equation, mais qu'il n'y a quelquefois aucune quantite, qui corresponde a celles qu'on imagine."

[58]"Et pourcequ'on ne trouue aucune racine, ny vraye, ny fausse, en ces deux dernieres Equations, on connoist de là que les quatre de l'Equation dont elles procedent sont imaginaires; ...".

[59](Footnote of Smith & Latham in Descartes [1954].) That is, all its roots are imaginary.

[60]"... & que le Problesme, pour lequel on l'a trouuée, est plan de sa nature; mais qu'il ne sçauroit en aucune façon estre construit, a cause que les quantités données ne peuuent se joindre; ...".

[61](Footnote of Smith & Latham in Descartes [1954].) That is, the given quantities cannot be taken together in the same problem.

[62]"Au reste il est a remarquer que cette façon d'exprimer la valeur des racines par le rapport qu'elles ont aux costés de certains cubes dont il n'y a que le contenu qu'on connoisse, n'est en rien plus intelligible, ny plus simple, que de les exprimer par le rapport qu'elles ont aux subtenduës de certains arcs, ou portions de cercles, dont le triple est donné. En sorte que toutes celles des Equations cubiques qui ne peuuent extre exprimées par les reigles de Cardan, le peuuent estre autant ou plus clairement par la façon icy proposée."

[63]See, for example, the beginning of Descartes' *Géometrie* ([1954]).

[64]For example Cardano [1968], p.12 and Bombelli [1572], p.246.

[65]Leibniz uses ⊓ for equality.

[66]"Je vous envoye le livre de Bombelli, dont je vous ay parlé. Vous y verrez (Bombelli [1572], p.225) comment il sse sert des racines imaginaires (il appelle par exemple $\sqrt{-121}$ ou $11\sqrt{-1}$ *piu di meno* 11; et $-\sqrt{-121}$ ou $-11\sqrt{-1}$ *mene di meno* 11) et comment il trouve par la la racine de l'equation 1^3 15^1 plus 4, c'est a dire y^3 ⊓ $15y + 5$. Il dit d'en avoir une demonstration en lignes, qu'il met aussi [1572], p.231), mais il y preuve seulement qu'une telle équation est possible, et que sa racine est quelque chose de reel, qui se peut donner en lignes. Mais il ne s'ensuit pas que l'operation par son *piu di meno* est bonne. Car quoyqu'il dise à la fin de la [p.226] que ces racines sont venues de l'équation, ce n'est par pourtant sans supposition. Il paroist aussi par la [p.226] qu'il ne pouvoit pas resoudre par cette methode l'équation y^3 ⊓ $12y + 9$, dont la racine rationelle est fausse ou negative, sçavoir -3. Il trouve neanmoins en esssayant par une autre methode (tirée aussi de Cardan) que l'equation se peut diviser par $y + 3$, ne sçachant pas que par cette même raison -3 en est la racine fausse: et il trouve par ce [autre] moyen la vraye $1\frac{1}{2} + \sqrt{5\frac{1}{4}}$, laquelle estant composée d'un nombre et d'une racine quarrée, ne pouvoit pas estre tirée des formules de Cardan, parceque les racines qu'on a par ces formules, sont toujours ou irrationelles cubiques ou nombres. D'où vient qu'il a cru que les formules de Cardan ne servent pas en cette rencontre, et ne sont pas generales."

[67]Cf. Birkhoff and MacLane [1941], p.450.

[68]And we now know there is no such radical solution for the general equation of the fifth degree (see, for example, Birkhoff and MacLane [1941], p.452 f.).

[69]"Ainsi je croy d'avoir demonstré le premier (1) que les formules de Cardan sont absolument bonnes et generales, soit extrahibles, soit non extrahibles, soit vrayes soit fausses ou negatives. (2) Que nous avons par ce moyen la resolution generale de toutes les equations cubiques. (3) J'ay trouvé le premier qu'on peut former des racines composees non extrahibles de tous les degrez pairs, qui contiennent des imaginaires et dont neanmoins la realité peut estre rendue palpable sans extraction, pour faire juger que la realité de telles formules n'est pas bornée par l'extrahibilité: dont l'exemple de

la formula $\sqrt{1+\sqrt{-3}} + \sqrt{1-\sqrt{-3}}$, qui vaut $\sqrt{6}$, est une preuve tres considerable. (4) Je demonstre, ce que personne a demonstré encor, que toute l'equation cubique, qui peut estre deprimée, contient une racine rationelle, pourveu que l'equation même soit proposée en termes rationaux. D'où il s'ensuit que celle qui ne peut estre divisée par l'inconnue + ou − un diviseur rationel du dernier terme, est solide. Proposition tres importante, puisqu'elle nous donne un moyen asseuré de scavoir si un probleme est solide en effect, ou s'il l'est seulement en apparance. Mr Descartes ne parle pas si positivement, car il dit, qu'il faut examiner toutes les quantités qui peuvent diviser le dernier, qu'il suppose estre un entier et rationel: et il semble qu'il n'ose pas dire toutes les nombres, ou toutes les quantités rationelles. De sorte qu'il nous laisse en doute, s'il ne faut pas aussi examiner les diviseurs irrationels: soit qu'il n'avoit point de demonstration assez convaincante pour les diviseurs rationels à l'exclusion des irrationels, soit qu'il n'ait negligé de parler plus exactement. De là vient aussi qu'on peut demonstrer en cinquieme lieu (5) par la seule analyse, sans aide de Geometrie, que toute l'equation cubique est possible, pourveu quelle soit conceue en termes possibles."

[70]But see McLennon [1923], p.317 f.

[71]Following Descartes, Leibniz denotes the absence of a square term by an asterisk.

[72]"Ne qua tamen [c]ausa dubitandi relinquatur, duplici demonstratione generali rem conficiemus, quae rationales irrationalesve non moretur. Prior demonstratio huc redit: Formula Cardanica satisfacit aequationi cubicae triradicali; omnis formula quae satisfacit aequationi cuidam, est ejus radix; ergo formula Cardanica est aequationis cubicae triradicalis radix. Omnis porro radix aequationis cubicae triradicalis est quantitas realis (ex hypothesi ideo enim triradicalem vocamus, quod tres habet radices reales, qualem illam esse, quae regulas Cardani respuere credebatur, dudum ostensum est; videatur inprimis Schotenius in appendice de aequationum cubicarum resolutione; neque vero plures quam tres habere potest radix cubica ulla). Ergo formula Cardanica (etiam tum cum ex cubica triradicali ducitur) est quantitas realis. Superest ergo tantum, ut ostendamus formulam Cardanicam etiam aequationi cubicae triradicali satisfacere, quod apparet, si in aequatione ejusmodi ut $x^3 * - qx - r \sqcap 0$ substituendo valorem ipsius x, nempe

$$A^3 \qquad + 3A^2B \qquad\qquad + 3AB^2 \qquad\qquad + B^3$$

$$x^3 \ \sqcap \ \frac{r}{2} + \sqrt{\frac{r^2}{4} - \frac{q^3}{27}} + \sqrt[q]{(3)\frac{r}{2} + \sqrt{\frac{r^2}{4} - \frac{q^3}{27}}} + \sqrt[q]{(3)\frac{r}{2} - \sqrt{\frac{r^2}{4} - \frac{q^3}{27}}} + \frac{r}{2} - \sqrt{\frac{r^2}{4} - \frac{q^3}{27}}$$

$$- qx \ \sqcap \qquad\qquad - \sqrt[q]{(3)\frac{r}{2} + \sqrt{\frac{r^2}{4} - \frac{q^3}{27}}} - \sqrt[q]{(3)\frac{r}{2} - \sqrt{\frac{r^2}{4} - \frac{q^3}{27}}}$$

$$- r \ \sqcap \ -\frac{r}{2} \qquad\qquad\qquad\qquad\qquad\qquad\qquad -\frac{r}{2}$$

$$\sqcap \ 0 \qquad \sqcap \ 0$$

Semper ergo formula Cardanica satisfacit, nec refert major minorve sit $\frac{q^3}{27}$ quam $\frac{r^2}{4}$."

[73] In the first formula involving square roots Leibniz has a footnote for the two occurrences of q : Nam $AB \sqcap q/3$ ut facile calculo ostendi potest. (For $AB = q/3$ as can easily be shown by calculation.) Further Leibniz uses $\sqrt{(3)}$ for $\sqrt[3]{}$.

[74] The last part of $\frac{r}{2} - \sqrt{\frac{r^2}{4} - \frac{q^3}{27}}$ is missing ·in Gerhardt's edition (Leibniz [1899]).

[75] La remarque que vous faites touchant des racines inextrahibles, et avec des quantitez imaginaires, qui pourtant adjoutées ensemble composent une quantité reelle, est surprenante et tout à fait nouvelle. L'on n'auroit jamais cru que $\sqrt{1+\sqrt{-3}} + \sqrt{1-\sqrt{-3}}$ fist $\sqrt{6}$, et il y a quelque chose de caché là dedans qui nous est incomprehensible."

Part 3 - Real Numbers

[1]This chapter and the next are revised versions of Crossley [1977–8] reproduced by kind permission of the Editor of the Australian Mathematical Gazette.

Chapter V - Irrationals

[1]Here we mean 1, 2, 3, ...; zero being specifically excluded.

[2]That is, numbers had been correlated with lengths.

[3]This tradition of hagiography continues: see Gorman [1979]. For a more considered view see Burkert [1972].

[4]Anatolius was bishop of Laodicea about A.D. 280 (Thomas [1939], vol.I, p.2.).

[5]Anatolius, cited by Heron, *Definitions* (ed. Heiberg), 160.8, translated in Thomas [1939], vol.I, p.3.

[6]However, Diogenes Laertius [1925], p.322, mentions several other teachers.

[7]Gow's reference to p.65 (of Friedlein's edition) appears to be incorrect. It should be 352 (see Proclus [1970], p.275.

[8](Footnote of T. Taylor.) "i.e. the priests of Jupiter".

[9]For example, we find in Neugebauer [1957], p.114: "The ephemerides alone are never a reliable source for the investigation of the basic empirical facts. At present it is completely impossible to write a "history" of Babylonian astronomy in its latest phase. All we do have is the ephemerides in a form excellently adapted to practical computation and to predicting new moons, eclipses, etc.".

[10]Neugebauer [1957], p.91, gives a list of 6 major items.

[11](Peet's footnote.) "[Problems] nos. 61B and 66 are perhaps the only exceptions."

[12](Gillings's footnote.) "H.W. Turnbull, *The Great Mathematicians*, 4th ed., Methuen, London (1951), p.2 f."

[13]Peet [1923], p.3 adds: "there is no reason to doubt the scribe's own statement that it was a copy of an older document [of the nineteenth century B.C.]."

[14]A mythical Egyptian king. See [b] in the References.

[15]However, Diogenes Laertius [1925], p.331, attributes it to Moeris who is reputed to have lived about the same time in Egypt (Smith [1861], vol. 2, p.109).

[16](Footnote of Neugebauer and Sachs.)   "From $\alpha_2 = \frac{1}{2}(\alpha_1 + \beta_1)$ and $\beta_2 = \frac{a}{\alpha_2}$ it follows that $\beta_2 = \frac{2a}{\alpha_1 + \beta_1} = \frac{2\alpha_1\beta_1}{\alpha_1 + \beta_1}$. This expression is known as the "harmonic mean" of α_1 and β_1 ."

[17](Footnote of Neugebauer and Sachs.) "More accurately: $\alpha = 1;24,51,10,35,...$"

[18](Footnote of Neugebauer and Sachs.) "It should be remarked that the expansion of $\sqrt{2}$ into a continued fraction also leads to (1), but not until the seventh step."

[19]See, for example, B. Abramenko [1958].

[20]Heath says in Euclid [1925], vol. III, p.2: "The actual method by which the Pythagoreans proved the incommensurability of $\sqrt{2}$ with unity was no doubt that referred to by Aristotle [Aristotle, *Prior Analytics* I.23, 41a,26-7] ... The proof formerly appeared in the texts of Euclid as X.117, but it is undoubtedly an interpolation."

Chapter VI - The Totality of Real Numbers

[1]Even rational numbers can be regarded as infinite decimals either by putting 0000 ... at the end or because the decimal recurs, e.g. $1/7 = .14285714285714 \ldots$.

[2]Davis and Hersh ([1981], p.147) coyly say:

> *The assertion has been made that mathematics is uniquely characterized by something known as "proof" but the assertion is a very Western one. Proof does not figure largely in the mathematics of the East.* (See Datta and Singh [1962] and Li Yan and Du Shiran [1987] *passim*.)

[3]Footnote of Knorr omitted (Knorr [1975], p.39).

[4]Robinson wrote the entry for Méray in the D.S.B..

[5]As an example where Archimedes' Axiom is not obeyed consider drops. One drop of water plus one drop of water may still be one drop of water, though of larger diameter.

[6]"Merito disputatur de numeris irrationalibus, an veri sunt numeri, an ficti. Quia enim in Geometricis figuris probandis, ubi nos rationales numeri destituunt, irrationales succedunt, probantque praecise ea, quae rationales numeri probare non poterant, certe ex demonstrationibus quas nobis exhibent: movemur & cogemur fateri, eos vere esse, videlicet ex effectibus eorum, quos sentimus esse reales, certos, atque constantes."

[7]"At alia movent nos ad diversam assertionem, ut videlicet cogamur negare, numeros irrationales esse numeros. Scilicet, ubi eos tentaverimus numerationi subjicere, atque numeris rationalibus proportionari, invenimus eos fugere perpetuo, ita ut nullus eorum in se ipso praecise apprehendi possit ... Non autem potest dici numerus verus, qui talis est ut praecisione careat, & ad numeros veros nullam cognitam habeat proportionalem. Sicut igitur infinitus numerus, non est numerus: sic irrationalis numerus non est verus numerus, quia latet sub quadram infinitatis nebula. Sitque non minus incerta proportio numeri irrationalis ad rationalem numerum, quam infiniti ad finitum."

[8]A binomium is a number of the form $a \pm \sqrt{b}$ where a^2/b^2 is rational but a/b is not.

[9]A medial line is a line whose square is equal to the product of lines commensurable in square but not directly. Thus, in modern dress, its length is of the form $\sqrt{ab}$ where a^2/b^2 is rational but a/b is not.

[10]"3. Circulus physicus est imago quaedam circuli mathematici. 4. Triangulus est polygoniarum omnium prima. 5. Omnium polygoniarum ultima est circulus. 6. Recte igitur describitur circulus mathematicus esse polygonia infinitorum laterum. 7. Mathematici itaque circuli circumferentia; nullum recipit numerum, necque rationalem necque irrationalem."

[11]"... hoc est, ipse circulus, videtur habere perimetrum infinitis gradibus incommensurabilem cum radio."

[12]"Aequales denique numeri definiri possent, que representant magnitudines eodem modo divisas (vel accuratius aliquovis simili modo)."

[13]It is not true in general that the solution of a cubic equation with rational or integer coefficients is expressible in terms of *real* radicals (cf. Birkhoff & MacLane [1941], p.450).

[14]"Als 3 $_1$ 7 $_2$ 5 $_3$ 9 $_4$, dat is te seggen 3 *Eersten*, 7 *Tweeden*, 5 *Derden*, 9 *Vierden*, ende soo mochtmen oneyndelick voortgaen. Maer om van hare weerde te segghen, soo is kennelick dat naer luyt deser Bepalinge, de voornoemde gethalen doen $\frac{3}{10}$, $\frac{7}{100}$, $\frac{5}{1000}$, $\frac{9}{10000}$, tsamen $\frac{3759}{10000}$."

[15]"Mais combien ce theoreme est veritable, toutesfois nous ne pouvons cognoistre par telle experience, l'incommensurance de deux grandeurs proposees; Premierement parce qu'a cause de l'erreur de noz yeux & mains (qui ne peuvent parfaictement veoir & partir) nous jugerions a la fin, que tous grandeurs tant incommensurables que commensurables, fussent commensurables. Au second, encore qu'il nous fust possible, de soubstraire par action, plusieurs cent mille fois la moindre grandeur de la majeure, & le côtinuer plusieurs milliers d'annes, toutesfois (estant les deux nombres proposez incommensurables) l'on travailleroit eternellement, demeurant tousiours ignorant, de ce qui à la fin en pourroit encore advenir; Ceste maniere donc de cognition n'est pas legitime, ains position de l'impossible, à fin d'ainsi aucunement declarer, ce qui consiste veritablement en la Nature; ceste incommensurance doncque est seulement notoire par les nombres incommensurables; ce que Euclide sçachant fort bien, aussi que tel invention d'incommensurabilité n'estoit suffisante pour ses propositions suyvantes (car sa dixiesme proposition enseigne trouver grandeurs incommensurables par le moyen des nombres) il l'a expliqué à la 8e proposition legitement selon les nombres, & ainsi le ferons nous en ceste premiere partie des definitions comme s'enfuit. *Definition I.* Grandeurs incommensurables sont celles, desquelles les nombres les explicans sont incommensurables."

[16]"De sorte qu'on peult ainsi infiniment approcher au vrai, mais iamais n'y peult on avenir par telle maniere: dont la raison est (...) que l'incommensurable ne peult estre commensurable."

[17]We omit theses V - VII.

[18]He does not, however, admit complex numbers as numbers. Cf. above, chapter IV, §3.

[19]"*These I.* Que l'unité est nombre. *These II.* Que nombres quelconques peuvent estre nombres quarrez, cubiques, de quart quantité, &c. *These III.* Que racine quelconque est nombre. *These IV.* Qu'il n'y a nombres absurds, irrationels, irreguliers, inexplicables, ou sourds."

[20]See for example Leibniz [1768], vol.III, p.406 where he says the sum of $1 - 1 + 1 - 1 + ...$ is 1/2.

[21]"Def. 1. Sit data infinita series a + b + c + d + e + f + g + h + &c. &

si successivae summae a, a + b, a + b + c, a + b + c + d, &c. continuo ad summam seriei vergunt, & ultimo propius accedunt, quam pro data quavis differentia, tum haec series dici possit convergens."

[22]A number is *transcendental* if it is not a root of any equation with rational (or equivalently integral or even surd) coefficients.

[23]Molk has a footnote which gives Méray [1869] as a reference.

[24]In fact Méray was the first to *publish* such an arithmetic treatment. Dedekind (see below, §7) claims to have thought of such a treatment in 1858.

[25]"*Ch. Méray* est le premier [see note 23] qui ait trouvé un sens purement arithmétique à l'expression *nombre irrationel*."

[26]I have not seen Méray's paper but the subject is treated in Méray [1894], vol. I, chapter I. The present account is based on this book and Molk's article (Molk [1904]).

[27]*...la condition nécessaire et suffisante pourqu'une variante donnée $\nu_{m,n,...}$ tende vers quelque limite commensurable ou incommensurable est qu'elle soit convergente, c'est-à-dire que la différence*

$$\nu_{m'',n'',...} - \nu_{m',n',...}$$

tende vers zéro quand les entiers $m', n', ..., m'', n'', ...$ augmentent tous indéfinement d'une manière quelconque.

[28](Cantor's footnote.) Thomas von Aquino, Opuscula XLII de natura generis, cap. 19 et 20; LII de natura loci; XXXII de natura materiae et de dimensionibus interminatis. Man vergleiche [= cf.]: C. Jourdain, *La Philosophie de Saint Thomas d'Aquin*, pag. 303-329; K. Werner, *Der heilige Thomas von Aquino* (Regensburg 1859), 2 Bd. p.177-201.

[29]"Der Begriff des "Kontinuums" hat in der Entwicklung der Wissenschaften überall nicht nur eine bedeutende Rolle gespielt, sondern auch stets die größten Meinungsverschiedenheiten und sogar heftige Streitigkeiten hervorgerufen. Dies liegt vielleicht daran, daß die ihm zugrunde liegende Idee in ihrer Erscheinung bei den Dissentierenden einen verschiedenen Inhalt aus dem Grunde angenommen hat, weil ihnen die genaue und vollständige Definition des Begriffs nicht überliefert worden war; vielleicht aber auch, und dies ist mir das wahrscheinlichste, ist die Idee des Kontinuums schon von denjenigen Griechen, welche sie zuerst gefaßt haben mögen, nicht mit der Klarheit und Vollständigkeit gedacht worden, welche erförderlich gewesen wäre, um die Möglichkeit verschiedener Auffassungen seitens der Nachfolger auszuschließen. So sehen wir, daß *Leukipp*, *Demokrit* und *Aristoteles* das Kontinuum als ein Kompositium betrachten, welches expartibus sine fine divisibilibus besteht, dagegen *Epikur* und *Lukretius* dasselbe aus ihren Atomen als endlichen Dingen zusammensetzen, woraus nachmals ein großer Streit unter den Philosophen entstanden ist, von denen einige dem *Aristoteles*, andere dem *Epikur* gefolgt sind; andere wieder statuierten, um diesem Streit fern zu bleiben, mit *Thomas von Aquino*, daß das Kontinuum weder aus unendlich vielen, noch aus einer endlichen Anzahl von Teilen, sondern aus *gar keinen* Teilen bestehe; diese letztere Meinung scheint mir

weniger eine Sacherklärung als das stillschweigende Bekenntnis zu enthalten, daß man der Sache nicht auf den Grund gekommen ist und es vorzieht, ihr vornehm aus dem Wege zu gehen. Hier sehen wir den *mittelalterlich-scholastischen Ursprung* einer Ansicht, die wir noch heutigentages vertreten finden, wonach das Kontinuum ein unzerlegbarer Begriff oder auch, wie andere sich ausdrücken, eine reine *apriorische* Anschauung sei, die kaum einer Bestimmung durch Begriffe zugänglich ware; jeder arithmetische Determinationsversuch dieses *Mysteriums* wird als ein unerlaubter Eingriff angesehen und mit gehörigen Nachdruck zurückgewiesen; schüchterne Naturen empfangen dabei den Eindruck, als ob es sich bei dem "Kontinuum" micht um einen *mathematisch-logischen Begriff*, sondern viel eher um ein *religiöses Dogma* handle. Mir liegt es sehr fern, diese Streitfrage wieder heraufzubeschwören, auch würde mir zu einer genaueren Besprechung derselben in diesem engen Rahmen der Raum fehlen; ich sehe mich nur verpflichtet, den Begriff des Kontinuums, so logisch-nüchtern wie ich ihn auffassen muß und in der Mannigfaltigkeitslehre ihn brauche, hier möglichst kurz und auch nur mit Rücksicht auf die *mathematische* Mengenlehre zu entwickeln. Diese Bearbeitung ist mir aus dem Grunde nicht leicht geworden, weil under den Mathematikern, auf deren Autorität ich mich gern berufe, kein einziger sich mit dem Kontinuum in dem Sinne genauer beschäftigt hat, wie ich es hier nötig habe."

[30]"... die Differenz $a_{n+m} - a_n$ mit wachsendem n unendlich klein wird, was auch die positive ganze Zahl m sei, oder mit anderen Worten, dass bei beliebig angenommemem (positiven, rationalem) ϵ eine ganze Zahl n_1 vorhanden is, so dass $|a_{n+m} - a_n| < \epsilon$, wenn $n \geq n_1$ und wenn m eine beliebige positive ganze Zahl ist. Diese Beschaffenheit der Reihe (1) drucke ich in den Worten aus: "*Die Reihe* (1) *hat eine bestimmte Grenze* b."."

[31]"... es lag mir nicht entgernt im Sinne, durch die Fundamentalreihen zweiter, dritter Ordnung etc. *neue* Zahlen einzuführen, die nicht schon durch die Fundamentalreihen erster Ordnung bestimmbar wären, sondern ich hatte nur die begrifflich verschiedene Form des Gegebens eins im Auge; es geht dies aus einzelnen Stellen meiner Arbeit selbst deutlich hervor."

[32](Footnote of Cantor.) "Es gehört also zu jeder Zahlengrösse ein bestimmter Punkt, einem Punkte kommen aber unzählig viele gleiche Zahlengrössen als Koordinaten in obigen Sinne zu; denn es folgt, wie schon angedeutet wurde, aus rein logischen Grunden, dass gleichen Zahlengrössen als Koordinaten *nicht* ein und derselbige Punkt zukommen kann". (Thus to every quantity there is a determined point but uncountably many equal quantities are assigned to one point as coordinates in the above sense; for it follows, as already indicated before, on purely logical grounds, that equal quantities *cannot* correspond to different points and one and the same point *cannot* be assigned as coordinate to unequal quantities.)

[33]"Hat diese Entfernung zur Masseinheit ein rationales Verhältnis, so wird sie durch ein Zahlengrösse des Gebietes A [the rationals] ausgedrückt; im andern Falle ist es, wenn der Punkt etwa durch eine Konstruktion *bekannt* ist, immer möglich, eine Reihe $a_1, a_2, \ldots, a_n, \ldots$ (1) anzugeben, welche die in §1 [as above] ausgedrückte Beschaffenheit und zur fraglichen Entfernung eine solche Beziehung hat, dass die Punkte der Geraden, denen die

Notes

Entfernungen $a_1, a_2, \ldots, a_n, \ldots$ zukommen, dem zu bestimmender Punkte mit wachsendem n unendlich nahe rücken. Dies drücken wir so aus, dass wir sagen: *Die Entfernung des zu bestimmenden Punktes von dem Punkte o ist gleich b, wo b die Reihe* (1) *entsprechende Zahlengrösse ist* ... [And similarly for higher order "Zahlengrösse".] Um aber den in diesem § dargelegten Zusammenhang der Gebiete der in §1 definierten Zahlengrössen mit der Geometrie der geraden Linie vollständig zu machen, ist nur ein *Axiom* hinzufügen, welches einfach darin besteht, dass auch umgekehrt zu jeder Zahlengrösse ein bestimmter Punkt der Geraden gehört, dessen Koordinate gleich ist jener Zahlengrösse, und zwar in dem Sinne gleich wie solches in diesem § erklärt wird. Ich nenne diesen Satz ein *Axiom*, weil es in seiner Natur liegt, nicht allgemein beweisbar zu sein. Durch ihn wird denn auch nachträglich für die Zahlengrössen eine gewisse Gegenständlichkeit gewonnen, von welcher sie jedoch ganz unabhängig sind."

[34]The actual mechanics in terms of what we now call "Cauchy sequences" was worked out in a tradition initiated by Méray [1869].

[35]See also Dauben [1970/71], p. 202-208 and Wilder [1973], p. 130.

[36]Footnote of Stolz omitted.

[37]"Die in dem Fundamentalsatze der Exhaustionsmethode liegende Voraussetzung, dass je zwei geometrische Grössen derselben Art vergleichbar seien ist theoretisch jedenfalls unzulässig; denn die analytische Geometrie vermag Curven nachzuweisen, welche keine endliche Länge haben. Hier giebt es keinen anderen Ausweg, als die Längen der Curven, die Flächenräume und körperlichen Inhalte als Grenzwerthe zu definieren."

[38]"... so denken wir uns durch die unbegrenzte Reihe $\phi_0, \phi_1, \ldots$ ein *neues von jeder rationalen Zahl verschiedenes Object* gesetzt ..."

[39]Cf. Dedekind [1963], p.6.

[40]Note also that corresponding to Méray's definition of equivalence, Dedekind needs to define equality of pairs of classes (A,B), (A',B'). This is done by saying (A,B) equals (A',B') if the class (or set) A has exactly the same members as A' and B has exactly the same members as B'.

[41]De Morgan remarked in a letter to Hamilton that he knew what an infinitely small quantity was no more than he knew what a straight line was (see vol.3, p.571-2 of R.P. Graves, *Life of Sir William Rowan Hamilton*, 3 vols., Dublin 1882-89, reprinted New York 1975); thus showing that he was aware of the problems of saying just what a line is.

[42]The Dedekind-Peano axioms also have this property, see chapter II, §6 above. There too the uniqueness is due to considering the totality of the numbers, in this case the natural numbers, which satisfy the axioms.

[43]For a discussion of set theory see Crossley et al. [1972] chapter V.

[44]This also applies to the natural numbers. See also the next footnote.

[45]To be precise, in the language of first order arithmetic with symbols for any relations or functions on the real numbers. See Robinson [1966] for full details.

[46]In 1960 Robinson told me that he regarded non-standard analysis as rehabilitating the infinitesimals which Leibniz and others had used. He never put to me the view, later expressed by others as Robinson's view, that Robinson's infinitesimals were *the same* as Leibniz's.

[47]This point has also been made in an unpublished paper by J. McGechie [1978].

[48]This point seems to be overlooked by Knorr [1975] (see for example his p.131, 170). It is a step so easy for us, yet we should bear in mind the number of years it took for complex numbers to be given an adequate geometric representation. Of course this remark is only an analogy. Hindsight often blinds us to awareness of the difficulties involved in the introduction of very simple new concepts.

[49]Thus Proclus [1970], p.28 writes: "Every man who knows his science of his art should make his arguments appropriate to the things with which he is dealing. ... Even in mathematics we cannot demand the same degree of accuracy in all parts."

[50]See also Davis and Hersch [1981], p.162.

Epilogue

[1]See also Wilder [1973], p.208-9.

References

Reference is made to the edition I have seen (unless otherwise noted). This gives rise to apparently odd dates, for example Proclus [1970], but when the date is read in conjunction with the bibliographical details the strangeness should disappear.

[a] *Dictionary of Scientific Biography*, ed. C.C. Gillispie, New York, 1970-1976). We refer to this as D.S.B..

[b] *Oxford Classical Dictionary*. ed. N.G.L. Hammond & H.H. Scullard, Oxford, second edition 1970.

[c] *The Anchor Bible*, vol. 38, Intro. &c. J. Massyngberde Ford, New York, 1975.

Abramenko, B. [1958]
> On Dimensionality and Continuity of Physical Space and Time. *Brit. J. Phil. Sci.* 34, p.95 f.

Aczel, P.H.G. [1977]
> An Introduction to Inductive Definitions, in Barwise, J. (ed.), *Handbook of Mathematical Logic*. Amsterdam, p.739-782.

Alfraganus [1537] = Ahmad ibn Muhammad ibn Kathïr,
> *Rudimenta Astronomica Alfragani .. Oratio introductoria in omnes Scientias Mathematicae Ioannis de Regiomonte, Patauij Labita ... .* Nuremberg.

Amir-Moez, A.R. [1961]
> A paper of Omar Khayyam, *Scripta Mathematica* 26, p.323-337.

Arnold, E. [1879]   *The Light of Asia*.   London.

Aulie, W.   [1957]   High-layered numerals in Chol (Mayan), *Int. J. of American Linguistics* 23, p.281-283.

Barrow, Isaac [1860]
> *The Mathematical Works of Isaac Barrow D.D..*   Ed. W. Whewell, Cambridge.

Barwise, J. (ed.) [1977]
> *Handbook of Mathematical Logic*.   Amsterdam.

Beman, W.W. [1897]   A Chapter in the History of Mathematics, *Proc. Amer. Assoc. for Advancement of Science* 46, p.33-50.

Beman, W.W. [1963]   *Essays on the Theory of Numbers*, J.W.R.Dedekind, tr. by W.W.Beman. New York.

198

Berlin, B. [1968]   *Tzeltal Numeral Classifiers.*   The Hague.

Berlin, B. and P. Kay [1969]
  Basic color terms.   Berkeley & Los Angeles.

Bernoulli, Jacob [1686]
  Excerpta ex iisdem litteris, *Acta Eruditorum* vol. V,
  p.360–361.

Billingsley, H. [1570]
  The Elements of Geometrie.   Tr. H. Billingsley, preface
  by J. Dee.   London.

Birkhoff, G. and S. MacLane [1941]
  A Survey of Modern Algebra.   New York.   (Our references
  are to the edition of 1953.)

Black, M. [1933]   *The Nature of Mathematics.* London.

Blackman, D.R. [1969]
  SI units in engineering.   Melbourne.

Blades, W. [1881]   *The Boke of St. Albans [Facsimile]*, with an introduction
  by W. Blades.   London.

Blagden, C.O., [1906]
  See Skeat, W.W. and C.O. Blagden.

Bodemann, E [1895]   *Die Leibniz-Handschriften der kön. öff. Bibliothek zu
  Hannover,* Hanover.

Bolzano, B. [1851]   *Paradoxien des Unendlichen.*   Leipzig.

Bombelli, R. [1572] *L'Algebra* (3 Books).   Bologna, reprinted with new title
  page   only   1579.   Complete   ed.   Bks.   I–V,   ed.
  E. Bortolloti.   Milan 1966.   (Our references are to
  this edition.)

Bortolotti, E. [1923]
  *Manoscritto   matematici,   risguardanti   la   storia   dell'
  Algebra   esistenti   nelle   Biblioteche   di   Bologna.*
  Catania.

Bortolotti, E. [1929]
  L'Algebra,   Opera   di   Rafael   Bombelli,   Cittadino
  Bolognese, *Archiv f. Geschichte d. Math.* 11,   p.407–424.

Bortolotti, E. [1922]
  Definizioni di numero, *Periodico di Mathematiche Serie
  IV,* vol. II, p.413–429.

Bortolotti, E. [1925]
  L'Algebra nella Scuola Matematica bolognese del secolo
  XVI, *Periodico di Matematiche* (4) 5, p.147–184.

References

Bortolotti, E. [1903–4]
    Lezioni di analisi algebrica.   Modena.

Bortolotti, E. [1924]
    La Storia della Matematica nella Università di Bologna.
    Bologna.

Bortolotti, E. [1923a]
    Origine e Primo Inizio del Calculo degli Immaginari,
    Scientia 33, p.385–394 (French tr. *ibid.*, suppl. 71–80).

Bosmans, M.H. [1926]
    La Théorie des Équations dans l'"Invention Nouvelle en
    L'Algèbre d'Albert Girard", *Mathesis* 40, p.59–67, 100–109
    and 145–155.

Brainerd, B. [1968]
    On the Syntax of Certain Classes of Numerical
    Expressions, in: Grammars for Number Names, *Foundations*
    of Language Supplementary Series, 7 , ed. H.B. Corstius.
    Dordrecht.

Brainerd, B. [1968a]
    A Transformational–Generative Grammar for Rumanian
    Numerical Expressions, in: Grammars for Number Names,
    Foundations of Language Supplementary Series, 7, ed. H.B.
    Corstius.   Dordrecht.

Brouwer, L.E.J. [1975]
    Collected Works, vol. I. Ed. A. Heyting. Amsterdam.

Burkert, W. [1972] *Lore and Science in Ancient Pythagoreanism.*   Cambridge,
    Mass.,   Eng. tr. by E.L. Minar, Jr. of *Weisheit und*
    Wissenschaft, Nuremberg 1962..

Bussey, W.H. [1917] The origin of mathematical induction, *American Math.*
    Monthly 24, p.199–207.

Cajori, F. [1909]   Review of Voss, A.   *Ueber das Wesen der Mathematik*,
    Leipzig & Berlin .   *Bull. Amer. M.S.* 15, p.405–409.

Cajori, F. [1918]   Origin of the name "Mathematical Induction", *American*
    Math. Monthly 25, p.197–201.

Cajori, F.[1901]   *A History of Mathematics*, New York.

Cantor, G. [1932]   *Gesammelte Abhandlungen*, ed. E. Zermelo.   Berlin.

Cantor, G. [1872]   Über die Ausdehnung eines Satzes aus Theorie der
    trigonometrischen Reihen, *Math. Annalen* 5, p.123–132 =
    Gesammelte Abhandlungen, ed. E. Zermelo, Berlin,
    p.92–102.

Cantor, G. [1879–1884]
Über unendliche lineare Punktmannigfaltigkeiten, *Math. Annalen* 15 (1879), p.1–7; 17 (1880), p.355–358; 20 (1882), p.113–121; 21 (1883), p.51–58 & 545–586; 23 (1884), p.453–488 = *Gesammelte Abhandlungen*, ed. E. Zermelo, Berlin (1932), p.139–246.

Cantor, M. [1965] *Vorlesungen über Geschichte der Mathematik,* second edition, volume II, reprinted New York.

Cardano, G. [1663] *Opera Omnia* (10 vols.), ed. C. Sponi, Leiden. Reprinted New York (1967).

Cardano, G. [1663] Sermo de plus et minus = *Opera Omnia*, vol. IV, p.435–439.

Cardano, G. [1968] *The Great Art or The Rules of Algebra*, tr. T.R. Witmer. Cambridge, Mass., U.S.A..

Cassirer, E. [1953] *The Philosophy of Symbolic Forms*. (Eng. tr.) 3 vols. New Haven.

Chadwick, J. [1973] *Documents in Mycenaean Greek*. Cambridge. (Second edition, first ed. by M. Ventris & J. Chadwick.)

Chadwick, J. [1976] *The Mycenaean World*. Cambridge.

Charleton, W. [1663]
Chorea Gigantum ... *Stone–heng*, ... *restored to the Danes*. London.

Chuquet, N. [1484] See Flegg[1985].

Codrington, R.H. [1885]
The Melanesian Languages. Oxford.

Colebrooke, H.T. [1817]
Algebra, with Arithmetic and Mensuration from the Sanscrit of Brahmegupta and Bháskara. London. Reprinted Walluf bei Wiesbaden (1973).

Corstius, H.B. (ed.) [1968]
Grammars for Number Names, Foundations of Language Supplementary Series, 7, Dordrecht.

Cossali, P. [1797] *Origine, trasporto in Italia ... dell' Algebra*, vol. I. Parma.

Crossley, J.N. [1977–8]
Number, I. Irrational Numbers, Gazette of the Aust. Math. Soc. 4 (1977), p.73–85 and 5 (1978), p.1–14.

Crossley, J.N. *et al.* [1972]
What is mathematical logic? Oxford University Press.

References

Datta, B. and A.N. Singh [1962]

> *History of Hindu Mathematics.* (Second edition; first edition 1935,8), Bombay.

Dauben, J.W. [1970/71]
The Trigonometric Background to Georg Cantor's Theory of Sets, *Archive for History of Exact Sciences* 7, p.181–216.

Davis, P.J. and Hersh, R. [1981]
> *The Mathematical Experience.* Boston.

Dawson, J. [1881] *Australian Aborigines.* Melbourne.

Dedekind, J.W.R. [1888]
> *Was sind und was sollen die Zahlen?* Brunswick. Tr. W.W. Beman in *Essays on the Theory of Numbers*, New York (1963). Also = Dedekind [1968], *Gesammelte mathematische Werke*, vol. III, p.335 f.

Dedekind, J.W.R. [1968]
> *Gesammelte mathematische Werke*, 3 vols. in 2. New York. Reprint of edition of 1930–2.

Dedekind, J.W.R. [1890]
> Letter to Keferstein. Tr. in van Heijenoort, J. (ed.) [1967], *From Frege to Gödel*, Cambridge, Mass., p.98–103.

Dedekind, J.W.R. [1912]

> *Stetigkeit und Irrationalenzahlen.* English tr. in Dedekind (1963).

Dedekind, J.W.R. [1963]
> *Essays on the Theory of Numbers.* tr. W.W. Beman. Dover, N.Y..

Dedron, P. and J. Itard [1974]
> *Mathematics and Mathematicians*, 2 vols., tr. J.V. Field. London.

Denvert, F.W. & W.H. Oakland [1968]
> Letter to Nature, *Nature* 220, p.311.

Descartes, R. [1954]
> *The geometry of René Descartes*, tr. D.E. Smith & M.L. Latham. Dover, N.Y..

Dieserud, J. [1908] *The Scope and Content of the Science of Anthropology.* Chicago.

Diogenes Laertius [1925]
　　　　　　　　Lives of the Eminent Philosophers, trans. R.D. Hicks, 2
　　　　　　　　vols.　London.　Reprinted 1959.

Diophantos [1893–95]
　　　　　　　　Diophanti Alexandrini Opera Omnia, 2 vols., ed. P.
　　　　　　　　Tannery. Leipzig.

Diophantos [1959]　*Diophante d'Alexandrie*, tr. into French & ed. P. ver
　　　　　　　　Eecke. Paris.

Diophantos [1890]　*Die Arithmetik und die Schrifte über Polygonzahlen des
　　　　　　　　Diophantus von Alexandria*, tr. &c. G. Wertheim. Leipzig.

Dobritzhoffer, M. [1784]
　　　　　　　　Historia de Abiponibus ... 3 vols.　Vienna.　English
　　　　　　　　tr. London 1882.

Dresner, S. [1971]　*Units of Measurement*.　Aylesbury.

Dummett, M.A.E. [1977]
　　　　　　　　Elements of Intuitionism.　Oxford.

Dumont D'Urville, J.S.C. [1830–34]
　　　　　　　　Voyage de la corvette l'Astrolabe ... *1826*, ..., *1829*.
　　　　　　　　13 vols, 4 atlases.　Paris.

Durkheim, E. and M. Mauss [1963]
　　　　　　　　Primitive Classification, tr. & ed. R. Needham. London.

van Egmond, W. [1977]
　　　　　　　　The Development of Algebra in Europe during the Later
　　　　　　　　Middle Ages (*Preprint of paper read at the XV Int. Cong.
　　　　　　　　of Hist. of Sci.*, Edinburgh, August 1977).

van Egmond, W. [1983]
　　　　　　　　The Algebra of Master Dardi of Pisa. *Historia
　　　　　　　　Mathematica*, 10, p.399–421.

Euclid [1883–1916]　*Euclides Opera Omnia* (10 vols.), ed. I.L. Heiberg & J.
　　　　　　　　Menge. Leipzig.

Euclid [1925]　　　*The thirteen books of Euclid's Elements*, tr. &c. T.L.
　　　　　　　　Heath. Cambridge. Second edition, reprinted New York
　　　　　　　　(1956).

Euclid [1883–1916]　*Data* = vol. VI of *Euclides Opera Omnia*, ed. Heiberg &
　　　　　　　　Menge, Leipzig.　German tr. *Die Data von Euklid* by C.
　　　　　　　　Thaer, Berlin (1962).

Euler, L. [1822]　　*The Elements of Algebra*, vol. I, tr. Rev. J. Hewlett
　　　　　　　　(3rd ed.).　London.

Eves, H. [1958]　　Omar Khayyam's solution of cubic equations, *The
　　　　　　　　Mathematics Teacher* 51, p.285–6.

References

Eves, H. [1969]   *An Introduction to the History of Mathematics*, third
                       edition, New York.

Feingold, M. [1984] *The Mathematicians' Apprenticeship*. Cambridge University
                       Press.

Fermat, P. [1891-1922]
                       Oeuvres de Fermat, ed. &c. P. Tannery & C. Henry. Paris.

Fibonacci   See Leonardo Pisano.

Flegg, G, C. Hay and B. Moss (eds.) [1985]
                       Nicholas Chuquet, Renaissance Mathematician. Dordrecht.

Fowler, D.H. [1979] Ratio in Early Greek Mathematics. *Bulletin of the
                       American Mathematical Society*, n.s. 1, p.807-846.

Fowler, D.H. [1980] Book II of Euclid's *Elements* and a pre-Eudoxan Theory of
                       Ratio, *Archive for History of Exact Sciences*, 22, p.5-36.

Fowler, D.H. [1981] Anthyphairetic Ratio and Eudoxan Proportion, *Archive for
                       History of Exact Sciences*, 24, p.69-72.

Frege, G. [1879]   *Begriffsschrift*. Halle.   Trans. in van Heijenoort,
                       (ed.), *From Frege to Gödel*, Cambridge, Mass.,
                       U.S.A. (1967), p.1-82.

Frege, G. [1884]   see Frege [1953].

Frege, G. [1953]   *The Foundations of Arithmetic*, Eng. tr. of *Die Grundlagen
                       der Arithmetik* of Breslau (1884) by J.L. Austin, (second
                       edition), Oxford.

Freudenthal, H. [1953]
                       Zur Geschichte der vollständigen Induktion, *Arch.
                       Internationales d'Histoire des Sciences* 6, p.17-37.

von Fritz, K. [1945]
                       The Discovery of Incommensurability by Hippasus of
                       Metapontum, *Annals of Maths*. 46, p.242-264.

Gamow, G. [1968]   Letter to Nature, *Nature* 219, p.765.

Gandz, S. [1932]   The Mishna ha Middot ... and the geometry of ...
                       al-Khowarizmi, ed. &c. S. Gandz, *Quellen u. Studien zur
                       Geschichte d. Math.*, *Astronomie u. Physik*, Abt. A2.
                       Berlin.

Geoffrey of Monmouth [1912]
                       Histories of the Kings of Britain.   (One ed. is
                       translated by S. Evans, London (1912).)

Gillings, R.J. [1972]
    Mathematics in the time of the Pharaohs. Cambridge,
    Mass., U.S.A.

Gödel, K. [1931] Über formal unentscheidbare Sätze der Principia
    Mathematica und verwandter Systeme I. *Monatshefte f.*
    Math. u. Physik 38, p.173-198. English tr. in van
    Heijenoort (ed.), *From Frege to Gödel*, Cambridge,
    Mass., U.S.A. (1967), p.596-61

Gow, J. [1884] *A Short History of Greek Mathematics.* Cambridge.
    Reprinted New York (1968).

Grant, E.B. (ed.) [1974]
    A Source Book in Mediaeval Science. Cambridge, Mass.,
    U.S.A.

Grattan-Guinness, I. [1977]
    Dear Russell - Dear Jourdain. London.

Hale, K. [1975] Gaps in Grammar and Culture, p.295-315 of *Linguistics and*
    Anthropology. In Honour of C.F. Vogelin, eds. M. Dale
    Kinkade, K.L. Hale & O. Werner. Lisse.

Halmos, P. [1986] I want to be a mathematician (Excerpts). *Math.*
    Intelligencer 8, p.26-32.

Hamilton, Sir W.R. [1967]
    The Mathematical Papers of Sir William Rowan Hamilton,
    vol. III, eds. H. Halberstam & R.E. Ingram. Cambridge.

Hands, R. [1975] *English Hawking and Hunting in the Boke of St. Albans.*
    Oxford.

Hardy, G.H. & E.M. Wright [1960]
    An introduction to the theory of numbers. (Fourth
    edition), Oxford.

Harriot, T. [1631] *Artis Analyticae Praxis*, ed. Aylesbury & Warner. London.

Haskins, C.M. [1957]
    The Renaissance of the twelfth century: Studies in the
    History of Mediaeval Science. New York.

Heath, T.L. (ed.) [1897]
    The Works of Archimedes. Cambridge. Reprinted New York
    (1912).

Heath, T.L. [1921] *A History of Greek Mathematics*, 2 vols. Oxford.
    Reprinted (1960).

Heath, T.L. [1885] *Diophantus of Alexandria, A Study in the History of Greek*
    Algebra. Cambridge. Second ed. 1910. Reprinted New York
    (1964).

References

Heath, T.L. [1925]   See also Euclid, *The thirteen books of Euclid's Elements*. (Second edition), Cambridge.   Reprinted New York [1956].

van Heijenoort, J. (ed.) [1967]
                    From Frege to Gödel.   Cambridge, Mass., U.S.A.

Hersh, R [1979]   Some   Proposals   for   Reviving   the   Philosophy   of   Mathematics. *Advances in Mathematics*, 31 p.31–50.

Hertz, R. [1960]   *Death and the right hand*, tr. R. & C. Needham.   London.

Heyting, A. [1956] Intuitionism. Amsterdam.

Hilbert, D. [1971] *Foundations   of   Geometry*,   second   edition.   Illinois. (Trans. of the 10th German edition.)

Houton La Billardière, J.J. [1800]
                    Relation du voyage à la recherche de La Perouse ... , 2 vols.. Paris.

Huxley, F.J. [1974] *The Way of the Sacred*.   New York.

Iamblichus [1894]   *In   Nicomachi   Arithmeticam   Introductionem*,   ed. H. Pistelli. Leipzig.

Iamblichus [1818]   *Life of Pythagoras (Vita Pythagorica)*, trans. T. Taylor. London.   Reprinted 1926.

Isidore of Seville [1911]
                    Isidori Hispalensis Episcopi Etymologiarum sive Originum libri   XX.   Recognovit   brevique   adnotatione   critica instruxit, W.M. Lindsay, 2 vols.   Oxford.

Jayawardene, S.A. [1965]
                    Rafael   Bombelli,   Engineer–Architect: Some   unpublished documents of the Apostolic Camera, *Isis* 56, p.298–306.

Jocano, F.L. [1969] Outline of Philippine Mythology.   Manila.

Johnstone, W.D. [1975]
                    For good measure.   New York.
Jordanus Nemorarius [1496]
                    Arithmetica (Iordani Nemorarii) decem libris demonstrata (per J. fabru Stapulescem). Paris.

Jowett, B. [1953]   *The Dialogues of Plato* (tr.), 4 vols.   (Fourth edition.) Oxford.

Jung, C.G. et al. [1964]
                    Man and his Symbols.   London.   Reprinted New York 1968. Reference is to the latter ed..

206

References

al-Karkhi [1853]   *Extrait du Fakhrî, traité d'Algèbre par Aboù Bekr Mohammed Nem Alhaçan Alkarhû*, tr. &c. F. Woepcke. Paris.

Katz, D. [1953]   *Animals and Men*. London. (Trans. by H. Steinberg and A. Summerfield from the German ed. of 1937.)

Keller, K.C. [1955] The Chontal (Mayan) Numeral System, *Int. J. of American Linguistics* 21, p.258–275.

Kennedy, H.C. [1973]
    Giuseppe Peano, Selected Works, tr., ed. &c. London.

Omar Khayyam [1851] *L'Algèbre d'Omar Alkhayyâmî*, tr. &c. F. Woepcke. Paris.

al-Khowarizmi, Mohammed ibn Musa [1831]
    Algebra, tr. F. Rosen. London. See also S. Gandz (ed. &c.), The Mishna ha Middot ... and the geometry of ... al-Khowarizmi, *Quellen u. Studien zur Geschichte d. Math., Astronomie u. Physik*, Abt. A2 (1932), Berlin.

Kline, M. [1972]   *Mathematical Thought from Ancient to Modern Times*. New York.

Knorr, W.B. [1975]   *The Evolution of the Euclidean Elements*. Dordrecht.

Knuth, D.E. [1968]   *The Art of Computer Programming*, vol. I. Reading, Mass., U.S.A..

Koestler, A. [1964] *The Act of Creation*. London.

Lakatos, I. [1976] *Proofs and Refutations*. Cambridge.

Lam, Lay-Yong and Shen, Kangsheng [1986]
    Mathematical Problems on Surveying in Ancient China. *Archive for History of Exact Sciences*, 36, p.1–20.

Lean, G. [1985–]   *Counting systems of Papua New Guinea*, 12 vols. (so far) plus Research Bibliography, Draft edition. Dept. of Mathematics, Papua New Guinea Institute of Technology, Lae, Papua New Guinea.

Leibniz, G.W. [1899]
    Der Briefwechsel von Gottfried Wilhelm Leibniz mit Mathematikern, ed. C.I. Gerhardt. Berlin. (Reprinted Hildesheim 1962.)

Leibniz, G.W. [1768]
    Opera Omnia ..., vol. III, ed. L. Duytens. Geneva.

Leonardo Pisano [1857, 1862]
    Scritti di Leonardo Pisano, 2 vols., ed. B. Boncompagni. Rome.

References

Lévy-Bruhl, L. [1951]
    Les Fonctions Mentales dans les Sociétés Inférieures.
    Paris (late ed.). (Eng. tr. by L.A. Clare, London
    [1926].)

Li Wenlin, [preprint]
    The Chinese indigenous tradition of mathematics prior to
    the introduction of modern mathematics in the 19th
    century. (Preprint.)

Li Yan and Du Shiran [1987]
    Chinese Mathematics: A Concise History, Translated by
    J.N.Crossley and A.W-C.Lun. Oxford.

Lohne, J.A. [1966] Dokumente zur Revalidierung von Thomas Harriot als
    Algebraiker, Archive for History of Exact Sciences 3,
    p.185-205.

Lovell, K. [1964] The Growth of Basic Mathematical and Scientific Concepts
    in Children. (Third edition.) London.

Lovett, R. [1899] The History of the London Missionary Society 1795-1895.
    London.

Marre, A. [1880]   Notice sur Nicolas Chuquet et son « Triparty en la
    science des nombres », Bulletino di Bibliografia e di
    Storia delle Scienze matematiche e Fisiche 13, p.555-659
    & 693-814.

Mariner, W.   See J. Martin [1818].

Marshack, A. [1970] Notation dans les gravures du paléolithique supérieur,
    Publications de l'Institut de Préhistoire de l'Université
    de Bordeaux, Mémoire No. 8, Bordeaux.   (Not seen by me.)

Marshack, A. [1972] The Roots of Civilization. London.

Martin, J. [1818] An account of the Natives of the Tonga Islands ...
    compiled and arranged from the extensive communications
    of Mr. William Mariner, 2 vols. (Second edition.) London.
    (Otherwise known as Mariner's Tonga Islands. Another ed.
    (1827), 2 vols. in 1, Edinburgh.)

Masotti, A. [1959] See Tartalea [1554].

Maurolico, F. [1575]
    Arithmeticorum libri duo, in Opuscula Mathematica,
    Venice.

McGechie, J. [1978] Unpublished paper.

McLennon, R.B. [1923]
    A contribution of Leibniz to the history of complex
    numbers, Amer. Math. Monthly 30, p.369-374.

208

Menninger, K. [1969]
  Number Words and Number Symbols, tr. P. Broneer. Cambridge, Mass., U.S.A..

Méray, C. [1869]   Remarques sur la nature des quantités définiés par la condition de servir de limites à des variables données, *Revue des Sociétés Savantes des Départements, Sci. Mat. Phys. Nat. (2)* 4, p.280-9.

Méray, C. [1894-8]   *Leçons nouvelles sur l'Analyse Infinitésimale et ses applications géometriques*, 4 vols., Gauthiers-Villars & fils. Paris.

Merrifield, W. [1968]
  Number Names in Four Languages in Mexico, in Corstius (ed.), Grammars for Number Names, *Foundations of Language Supplementary Series*, 7, Dordrecht, p.91-102.

Michell, J. [1973]   *City of Revelation*. London.

Molk, J. [1904]   Nombres irrationels et notion de limite (Exposé, d'après l'article allemand de A. PRINGSHEIM), in *Encyclopédie des Sciences Mathétiques pures et appliquées*, ed. J. Molk, vol.I, p. 133-208, Paris.

al-Nadīm, Ibn [1871-2]
  Fihrist, ed J. Roediger & A. Mueller, 2 vols.   Leipzig.

Nemorarius [1496]   See Jordanus Neomorarius.

Neugebauer, O. [1935]
  Mathematische Keilschrift-Texte.   Berlin.   Reprinted Berlin 1973.

Neugebauer, O. and A. Sachs [1945]
  Mathematical Cuneiform Texts.   New Haven, Conn., U.S.A..

Neugebauer, O. [1957]
  The Exact Sciences in Antiquity.   Providence, R.I., U.S.A. (Second edition.) Reprinted New York 1962.

Newton, I. [1972]   *The Mathematical Papers of Isaac Newton*, vol. V, 1683-1684, ed. D.T. Whiteside. Cambridge.

Nicomachus [1926]   *Nicomachus of Gerasa. Introduction to Arithmetic*, trans. M.L. D'Ooge with studies in Greek arithmetic by F.E. Robbins & L.C. Karpinski. New York. Reprinted 1972.

Novy, L. [1973]   *Origins of Modern Algebra*. Groningen.

References

Oughtred, W. [1648] *Clavis Mathematicae*, trans..   London 1694.

Pacioli, Fra. Luca [1494]
      *Suma   de   Arithmetica   Geometria   Proportioni   &
      Proportionalita*. Venice.

Paris, J. & L. Harrington [1977]
      A mathematical incompleteness in Peano Arithmetic, in
      Barwise   (ed.),   *Handbook   of   Mathematical   Logic*.
      p.1133-1142. Amsterdam.

Pascal, B. [1665]   *Traité du triangle arithmétique*. Paris.

Pascal, B. [1963]   *Oeuvres complètes*, ed. L. Lafuma. Paris.

Patrick, J.D. and C.S. Wallace [1982]
      Stone circle geometries: an information theory approach.
      p.231-264. D. Heggie (ed.). *Archaeoastronomy in the Old
      World*. Cambridge.

Peano, G. [1891]   Sul concetta di numero, *Rivista di Matematica 1*,
      p.87-102.

Peano, G. [1908]   *Formulario Mathematico Editio V (Tomo V de Formulario
      completo)*. Turin.

Peet, T.E.   [1923]   *The Rhind Mathematical Papyrus*.   London.   Reprinted
      Neudeln, Liechtenstein 1970.

Philip, J.A. [1966] *Pythagoras and Early Pythagoreanism*. Toronto.

Piaget, J. & A. Szeminska [1941]
      La genèse du nombre chez l'enfant. Neuchâtel.   Eng. tr.
      by C. Gattegno & F.M. Hodgson, London 1952.

Poincaré, H. [1905] *Science and Hypothesis*.   London.   Reprinted New York
      1952.

Porphyry [1815-16] *Porphyrii de Vita Pythagorae Liber*, p.2-123 of part II
      of *Iamblichi Chalcidensis ex Coele-Syria De Vita
      Pythagorica Liber*, ed. M. Th. Kriessling. Leipzig.

Proclus [1970]   *A commentary on the first book of Euclid's Elements*,
      trans. G.R. Morrow.   Princeton, N.J., U.S.A..

Rabinovitch, N.L. [1969]
      Rabbi Levi Ben Gershon and the Origins of Mathematical
      Induction.   *Archive for History of Exact Sciences* 6,
      p.237-248.

Rashed, R. [1972]   L'induction mathématique: al-Karajī, as-Samaw'al, *Archive
      for History of Exact Sciences* 9, p.1-21.

Rashed, R. [1974,1975]
Les travaux perdus de Diophante (I), (II), *Revue d'histoire des sciences* XXVII, p.97-122, and XXVIII, p.3-30.

Rashed, R. (ed.) [1975]
L'art de l'algèbre de Diophante trad. en arabe par Qusṭa ibn Luqa. Cairo.

Rigaud, S.P. (ed.) [1832]
Miscellaneous works and correspondence of Dr. Bradley. Oxford. With supplement, Oxford 1833.

Robbins, F.E. and L.C. Karpinski
See Nicomachus [1926].

Robinson, Abraham [1966]
Non-standard analysis. Amsterdam.

La Roche, E. (Villefranche) [1520]
Larismetique nouvellement composee par maistre Estienne de la roche dict Villefrãche. Lyon.

van Rootselaar, B. [1962-65]
Bolzano's Theory of Real Numbers. *Archive for History of Exact Sciences* 2, p.168-180.

Rose, P.L. [1976] *The Italian Renaissance of Mathematics.* Geneva.

Rosen, F. [1831] *The Algebra of Mohammed ben Musa.* London.

Rosenfeld, A. [1971]
Review of Marshack, Notation dans les gravures du paléolithique supérieur, in *Antiquity* 45, p.317-319.

Ross, W.D. (ed.), [1908-52]
Aristoteles: Works. (12 vols.) Oxford.

Roth, W.E. [1908] Counting and enumeration, *Records of the Australian Museum, Sydney* 7, p.79-82.

Russell, B.A.W. [1911]
Sur les axiomes de l'infini et du transfini, *Comptes Rendues des séances de la Soc. Math. de la France* 2, p.22-35 = *Bull. Soc. Math. France* 39 (1911) (reprinted 1967), p.488-501, tr. I. Grattan-Guinness in *Dear Russell - Dear Jourdain*, London (1977).

Saidan, A.S. [1966] The Earliest Extant Arabic Arithmetic, Kitab al-Fusul fi al Hisab al-Hindi of Abu al-Hasan, Ahmad ibn Ibrahim al-Uqlidisi. *Isis* 57, p.475-490.

References

Saidan, A.S. [1978] *The arithmetic of al-Uqlīdisī:* translated and annotated by A.S.Saidan. Dordrecht.

Schebesta, Paul [1928]
Among the Forest Dwarfs of Malaya. Kuala Lumpur. Reprinted 1973.

Schürmann, C.W. [1844]
A vocabulary of the Parnkalla Language ... . Adelaide.

Scriba, C.J. [1968] *The Concept of Number.* Mannheim–Zürich.

Seidenberg, A. [1960]
The Diffusion of Counting Practices, *Univ. of California Publications in Mathematics* 3, p.215–300.

Seidenberg, A. [1962]
The ritual origin of counting, *Archive for History of Exact Sciences* 2 (1962–6), p.1–40.

Sesiano, J. [1982] *Books IV to VII of Diophantos'Arithmetica in the Arabic translation attributed to Qustā ibn Luqā.* Berlin.

Sieveking, A. [1972]
Review of Marshack, *The Roots of Civilization,* in *Antiquity* 46, p.329–330.

Sinaceur, Md.-Allal [1976]
Sur l'algèbre de Diophante. Revue de Rashed, Les travaux perdus de Diophante (I), (II), in *Revue d'Histoire des Sciences* 39 (1936), p.167–170.

Skeat, W.W. & C.O. Blagden [1906]
Pagan races of the Malay Peninsula, 2 vols. London.

Skolem, T. [1934] Über die Nicht-characterisierbarkeit der Zahlenreihe mittels endlich oder unendlich vieler Aussagen mit ausschliesslich Zahlenvariablen, *Fund. Math.* 23, p.150–161.

Smith, D.E. [1923] *History of Mathematics,* vol. I. New York.

Smith, D.E. [1959] *A Source Book in Mathematics,* vol. I, reprinted, New York.

Smith, G.C. [undated]
Leibniz's investigation of the laws of algebra, *Monash University Mathematics Department, History of Mathematics Paper no. 2.*

Smith, W. [1861] *Dictionary of Greek and Roman Myth.* vol. II, London.

Smorynski, C. [1983]
    "Big" News from Archimedes to Friedman, *Notices of the American Mathematical Society* 30, p.251-256.

Stevens, H. [1900] *Thomas Hariot and his Associates*. London.

Stevin, S. [1635] *L'Arithmétique* ..., Leyden.

Stevin, S. [1958] *The Principal Works of Simon Stevin* (3 vols.), ed. D.J. Struik. Amsterdam.

Stevin, S. [1585] *De Thiende*. Leyden.

Stifel, M. [1544] *Arithmetica Integra*. Nuremberg.

Stolz, O. [1885] *Vorlesungen über Allgemeine Arithmetik. Erste Theil*. Leipzig.

Strehlow, T.G.H. [1944]
    Aranda Phonetics and Grammar, in *The Oceania Linguistic Monographs* 7, p.101-5.

Struik, D.J. [1958] Omar Khayyam, mathematician, *The Mathematics Teacher* 51, p.280-285.

Struik, D.J. [1969] *A Source Book in Mathematics, 1200-1800*, Cambridge, Mass., U.S.A..

Tanner,R.C.H. [1967]
    Thomas Harriot as Mathematician, *Physis* 9, p.235-292.

Tartalea, N. [1554] *Quesiti et Inventioni diverse de Nicolo Tartalea. Brisciano.* Reprinted with Intro. by A. Masotti, Brescia 1959.

Thom, A. [1967] *Megalithic Sites in Britain*. Oxford.

Thomas, Ivor (trans.) [1939]
    Greek Mathematics, 2 vols., London. Reprinted 1957.

Thomson, W.R. [1975]
    Friars in the Cathedral. Toronto.

Thorpe, W.H. [1966] *Learning and Instinct in Animals*. London.

Turnbull, H.W. [1951]
    The Great Mathematicians, 4th ed., London.

Tylor, E.B. [1871] *Primitive Culture*, 2 vols. London. (Partly reprinted as *The Origin of Culture*, New York 1958.)

Vacca, G. [1909] Maurolycus, the first discoverer of the principle of mathematical induction, *Bull. Amer. Math. Soc.* 16, p.70-73.

References

Vacca, G. [1910]   Sulla storia del Principio d'Induzione Complete, *Bolletino di Bibliografia e Storia delle Sci. Mathematica* 12, p.33-35.

Vaillant, G.C. [1944]
The Aztecs of Mexico. New Mexico.   Another ed. London 1950.

Viète, F. [1646]   *Opera Mathematica.* Leiden.   Reprinted Hildesheim 1970.

de Vogel, C.J. [1966]
Pythagoras and Early Pythagoreanism. Assen.

Voss, A. [1908]   *Ueber das Wesen der Mathematik.* Leipzig & Berlin.   (Not seen by me.)

van der Waerden, B.L. [1975]
Science Awakening, vol. I. (Fourth edition.)   Groningen.

Wagner, D.B. [1979] An early Chinese derivation of the volume of a pyramid: Liu Hui, Third century A.D.. *Historia Mathematica* 6, p.164-188.

Wallis, J. [1699]   *Opera Mathematica,* 3 vols.   Oxford. Reprinted Hildesheim [1972].

Wallis, J. [1684]   *A Treatise of Algebra.* Oxford (1685, actually 1684). Latin ed. in *Opera Mathematica* vol. II, Oxford 1699, reprinted Hildesheim 1972.

Wang, Hao [1974]   *From Mathematics to Philosophy.* London.

Waring, E. [1776]   *Meditationes Analyticae.* Cambridge.

Weber, H. [1891-2]   Leopold Kronecker [obituary], *Jahresbericht der Deutschen Mathematiker-Vereinigung* 2, p.5-31.

Wertheim, G. [1890] *Die Arithmetik und die Schrift über Polygonzahlen des Diophantus von Alexandria,* tr. &c..   Leipzig.

Wieleitner, H. [1927]
Zur Frühgeschichte des Imaginären, *Jahresbericht der Deutsche Math. Vereinigung* 36, p.74-88.

Wilder, R.L. [1965] *Introduction to the Foundations of Mathematics,* second edition. New York.

Wilder, R.L. [1973] *Evolution of Mathematical Concepts,* second edition. Falmouth.

Wittgenstein, L. [1958]
Philosophical Investigations. Oxford.

References

Woepcke, F. [1853] *Extrait du Fakhrī, Traité d'algèbre par Abou Bekr Mohammed ben Alhaçan Alkharkhī.* Paris.

Zacher, H.J. [1973] *Die Hauptschrifte zur Dyadik von G.-W. Leibniz.* Frankfurt am Main.

Index

Index

Index

Faultes escaped

Marvaile not (gentle reader) that faultes ...
have escaped in the correction of this booke. ...

And if you happen in reading to finde any more
faultes ... I trust you will therefore impute no blame
either unto me or to the Printer, but gently amend and
correct them accepting our good mind, which was to have
had the booke passed into your handes utterly without
fault, as touching the Printing.

H. Billingsley, *The Elements of Geometrie ... .*
London (1570), f.463 v.